# CHRISTIANITY

THE

# LOGIC OF CREATION

BY

## HENRY JAMES

AUTHOR OF

"THE CHURCH OF CHRIST NOT AN ECCLESIASTICISM," ETC.

" Felix qui potuit rerum cognoscere causas,
Atque metus omnes et inexorabile fatum
Subjecit pedibus, strepitumque Acherontis avari."

VIRGIL.

## LONDON
WILLIAM WHITE, 36 BLOOMSBURY STREET

1857

# PREFACE.

THE following Letters were actually written to a
friend in London, and are published at his suggestion. They are themselves but a preface to a larger
discourse upon the same theme, which the writer
hopes some day to accomplish. Christianity was
the revelation of an utterly unsuspected life of God
within the strictest limits of human nature; and,
like all true revelation of spiritual things, was an
inverse form of its own interior substance. For
this is the distinction between revelation, properly
so called, and information, that the one constitutes
an inverted image of the truth, the other a direct
image; or that the one is symbolic and speaks
mainly to the soul, while the other is purely sta-
tistical and addresses chiefly the senses. Hence
it is that revelation has always shielded and fostered human freedom, while mere information is
sure to crush it. None of the sects exhibit so
servile a temper as those who pretend to the most
authoritative information about spiritual things.
Look at the Swedenborgians, for example. And
*Mediumship*, as it is called, is growing to be the
aspiration and profession of thousands, who are

state of the case. Viewed literally, the Lord was
an historic person, the most finite and dejected of
men. Viewed spiritually, however, he is the life
of universal man, existing nowhere but in the
individual soul conjoined with God. To the spiritual apprehension the Lord is not a finite historic person, capable of being outwardly discriminated from other persons : He is the infinite Divine
love and wisdom in union with every soul of man.
*He has no existence or personality apart from such
union.* You Swedenborgians are wont to talk of
the glorification of the human nature in the Christ,
as of certain phenomena which transpired within
the *spatial* limits of Christ's body, and remain
permanently confined to those limits throughout
eternity, thus practically turning the Christ into a
mere miracle, or Divine *tour de force*, fit for Barnum's museum of curiosities. I am persuaded
that nothing more baldly sensual exists out of
Heathendom, than much of this prevalent orthodox lore. Swedenborg tells us with all his might
that *time, space* and *person,* are unreal existences :
that real existence is of an intensely human quality, being made up exclusively of affection, and of
thought derived from that affection : and yet his
reputed followers go on to cogitate the spiritual
world as compounded of space, time, and person,
precisely as if he had never uttered a word upon
the subject. " *Not any person,*" says Swedenborg,
" named in the Word *is perceived in heaven,* but

instead thereof the thing which is represented by that person."—*A. C.*, 5225. "There are three things," he says, "which perish from the literal sense of the Word, while the spiritual sense is evolving, namely whatsoever pertains to *time*, to *space*, and to *person*. The conception of time and space perishes, because these things are peculiar to nature, and spiritual thought is not determined to person, because a view to person in discourse contracts or limits the thought, and doth not render it unlimited: whereas what is extended and unlimited in discourse gives it universality, and fits it to express things innumerable and ineffable. Angelic discourse, especially that of the celestial angels, is of this character, being comparatively unlimited, and hence it connects itself with the infinite and eternal, or the *Divine* of the Lord."—*A. C.*, 5253. See also 5287, 5434.

Yours truly,

———.

# LETTER II.

*Paris, Sept. 29th*, 1856.

MY DEAR W.

You ask me to be a little more explicit in stating my views of New Church truth. I am not aware that there is anything recondite in my views. Ever since I knew Swedenborg's books I have of course been put upon my guard against my naturally sensuous and irrational views of creation. No doubt one learns wisdom slowly, but I may truly say that I no longer incline to regard creation as a physical act of God, and have ceased attributing to Him material modes of being. I feel, indeed, a hearty disrelish of the popular cant which, while professing to maintain the *spiritual* contents of the Scriptures, perpetually degrades the Divine creation, redemption, and providence, into mere historic problems like the French Revolution or the Battle of Waterloo. If the design of the New Testament be to give us historical information, no book was ever more undivinely constructed. *Robinson Crusoe* is a masterpiece of skill beside it, and the American *spūks* and table-

tippers, though their talk be only of the dreariest
millinery and mud of things, are yet more luminous than evangelists and apostles. No doubt all
spiritual truth falls at last into the historic plane,
in order that it may become cognizable in that
disguise to unspiritual or natural eyes. And the
Divine creation, redemption, and providence, obey
of course this universal law. But how base must
we deem the intelligence, which insists upon viewing the spiritual truth as identical with its historic
ultimation! How base, in other words, must we
consider that spiritual state which regards a *Divine*
operation as observing *strictly personal limits, or
shutting itself up to the experience of an individual
bosom.* Thus I have often been checked, in speaking of the Incarnation as a scientific verity, by the
suggestion that "the Incarnation took place only
in the Christ, and could be true therefore only of
his experience."

But those who talk in this way, under the impression that they are honouring the Lord, might
much more profitably employ their energies in
"whistling jigs to a milestone." Depend upon it,
the milestone will up and dance, long before any
angel will be caught in that foolish trap. It is a
trap, and nothing more. The idea is that in honouring Christ personally we honour Him spiritually, and so shall get to be honoured by Him.
There is no persuasion more puerile. We honour
Christ spiritually only by forgetting every personal

and limitary fact about Him, or rather by seeing
in these facts only their universal spiritual meaning, the meaning they reflect upon universal man
in relation to God. All the literal facts—Christ's
life, death, and resurrection,—are unspeakably precious—why? Because they contain some *magical*
virtue? Assuredly not, but only because they
*reveal* a truth which they do not constitute, a
truth which relates universal man to God. Spiritual Christianity drops out the carnal Jesus, or
no longer sees Christ after the flesh. It drops the
man born of the virgin Mary, six feet high more
or less, of an uncomely aspect, bent and seamed
with sorrow, to see henceforth the glorified or Divine Man who is the intimate and omnipresent
secret of creation. Spiritually viewed, Christ is
the inmost and vital selfhood of every individual
bosom, bond or free, rich or poor, good or evil,
whether such bosom be reflectively conscious of
the truth or not. But in saying this I should be
very sorry to be understood as saying, that the
literal Man Jesus of Nazareth becomes lifted out
of His native environment, and personally inserted
in every individual bosom. This would be too
absurd. What then do I mean? I mean simply
to indicate the spiritual significance of the Christ.
I mean to say that the birth, life, death and glorification of Christ spiritually imply, that *infinite
love and wisdom constitute the inmost and inseparable life of man, and will ultimately vindicate*

*their creative presence and power by bringing the most degraded and contemned forms of humanity into rapturous conscious conjunction with them.* When I think spiritually of the Christian truth, I do not think of Jesus personally, except as it were to anchor or define my thought. I think quite away from Him personally indeed, and fix my attention upon what is universal to man, or upon the life of universal human fellowship which the Divine love is now engendering in your bosom and mine, and that of all other men, by the stupendous ministry of science. The Christian facts attest, reveal, predict this universal spiritual life of man, this redemption of the natural mind, because they are a real ultimation of it. Every incident of Christ's personal history grew out of this unseen and unknown Divine operation in humanity, and were thus a mystical and endless revelation of it, such a revelation as human intelligence permitted. There could have been no scientific *information* upon the subject of course, because no angel even knew the wonders of the Divine love implied in the intimacy of His conjunction with human nature. By the very necessity of the case, therefore, the great and inscrutable truth could only look forth under a veil, and wait for the gradual unfolding of human reason to be discerned in its just spiritual proportions. That just discernment is now taking place. Men are everywhere now beginning to drop the tedious cant of mere *per-*

*sonal* homage to Christ, and insist upon finding a universal humanitary meaning in His truth, a meaning which shall vitally associate with God every man of woman born, whatever be his natural limitations and infirmities.

Thus the Divine Incarnation is with me a *spiritual* truth before—or, in order to its—becoming a *natural* one. I value the natural facts only because they contain something higher and better than themselves, something which relates you and me and all mankind to the inmost and exhaustless heart of God. The entire history of the church from Adam to Christ inclusive, is only a series of effects from *a real Divine operation in the spiritual world, which is the universal mind of man;* and your and my spiritual experience with that of our remotest natural descendants, constitute the substance of that world, quite as much as does that of Moses or David. The spiritual world, or the mind of man, is out of space and time; and all God's alleged spiritual judgments which were expressed or ultimated in the life of Christ, claim your and my bosom for their veritable ground or arena, quite as much as they do that of any one who died before Christ. Thus we can spiritually understand Christianity only in so far as we rightly apprehend the life which is taking place in our own bosoms and that of our contemporaries. All the Swedenborgs who ever lived will not avail us here, but only the clear and reverential insight into what God is now effecting in the universal mind of man. I for my part see, very clearly, that God is begetting by the ministry of science such a recognition of human society, fellowship, or equality in the bosom of man, as that bosom has never conceived, much less known, and can never again lose sight of: such a recognition, indeed, as must ere long prostrate every throne and altar now erected upon the twin dogmas of human inequality and depravity, and by means of such prostration bring the whole disunited family of man into conditions of mutual knowledge, love and reverence. And seeing this, I see that such and no less is the spritual force of Christianity: that this boundless blessing of God upon man's natural life, and by means of that upon his spiritual life, is the great and universal burden of the Christian letter, and I consequently value that letter not with any servile estimation, but with the hearty relish of one who has tasted its endless and ineffable spiritual contents.

Yours truly,

————.

# LETTER III.

*Paris, Oct.* 1.

MY DEAR W.

THE great disease of the religious mind at present is, that it obstinately persists in looking upon religion as a *private* question instead of a *public* one, as an affair of the individual conscience instead of the associated one.   One is not surprised at the old sects continuing in this traditional way, but I am surprised that you, who read Swedenborg, should not have begun to get out of it, for Swedenborg shews us in every page of his books, that revelation proceeds 'upon strictly *universal* principles, and that not one single word of it is to be spiritually interpreted in a private or personal sense.

The old theory of religion is that God is a respecter of persons, that He approves one sort, the morally good, and saves them; and disapproves another sort, the morally evil, and damns them.   Viewed spiritually, of course this is arrant superstition, because all men are alike worthless in the Divine sight, the morally good and the

morally evil; and God would quite as gladly,
therefore, bless one as the other, only that the
morally good man, in consequence of the conceit
he derives from the general estimation in which
he is held, will not permit himself to be blessed.
These are they, who being secure of the honour
that comes from men, do not aspire after that
which comes solely from God. " It is easier for
a camel to go through a needle's eye, than for a
*rich* man to enter the Divine kingdom." And
Swedenborg shews you that no angel in heaven
ever feels himself rich in comparison with others,
without, *ipso facto*, tumbling into infernal company. Still the church has maintained itself in
the world hitherto on this most sandy foundation.
It has always been thought that there was an
*essential* difference between the saint and the
sinner, and that the distinction between heaven
and hell was measured by that difference: so that
practically what we have all sought to do in order
to merit heaven has been, to make ourselves different from certain other people, whom the world
cheerfully consigns to hell. In this manner the
ecclesiastical temper has proved to be one of
intense Pharisaism and self-righteousness, filling
the world of spirits with all manner of flatulent
falsity and obstruction. The last judgment, as
Swedenborg proves, had exclusive reference to
these pestilent and cruel moralists. See his little
book on the *Last Judgment*, 69, with the *Con-*

*tinuation*, 10, 16 : and also the *Doctrine of Faith of the New Jerusalem*, 64.

No intelligent reader of Swedenborg ought to fail to perceive how low this temper is, and how utterly repugnant to the genius of the spiritual dispensation. According to Swedenborg, there is no *essential* difference between saint and sinner, angel and devil; in fact, there is no actual difference, even, save what is made by the one acknowledging the Lord and the other not doing so. According to the unvarying testimony of this enlightened man, it is the Lord alone who, by His presence inwardly in the angel, causes him to be an angel, and by His absence inwardly from the devil, suffers him to be a devil. Now what does Swedenborg mean when he speaks in this way? What does he mean by the LORD who is inwardly present or absent from man? Does he mean a literal person, capable of outward or sensible discrimination from other persons? Surely this would be ridiculous, for no man nor angel could possibly exist with an additional person to himself included in his own skin. By the Lord regarded spiritually or rationally, then, we do not mean any literal or personal man, capable of being sensibly comprehended; but we mean that Divine and universal life in man, which grows out of the conjunction of the infinite Divine Love with our finite natural loves, and which was perfectly manifested and ultimated in the Christ, considered as

the end of the old or carnal economy, and the
beginning of the new or spiritual one. Christ
was conceived of the Holy Spirit, born of the
virgin mother, lived, died, and rose again, only
by virtue of this latent Divine life in humanity
waiting to develop itself in the fulness of time,
and seeking meanwhile to give itself intellectual
anchorage or projection by all those literal facts.
Had there not been a realm of life stored away in
the still unsunned depths of human nature, unperverted by human folly, unstained by human
sin, a realm of life in which the Divine Love
reigns supreme, attracting the cordial, and glad,
and boundless homage even of our most natural and selfish loves, then Christianity must
have proved an illusion and abortion. For the
glorification of the Christ, which is its great truth,
obviously pre-supposes the *spontaneous* subjection
of self-love to charity, or hell to heaven, and of
heaven to the Divine : and when self-love is *spontaneously* subject to brotherly love, human nature
is redeemed, and every man becomes thenceforth
*naturally conjoined* with God. I do not say that
every one thereby becomes *spiritually regenerate*,
for spiritual regeneration, or new birth, implies
the existing disjunction of the Divine and human
natures, and has never taken place except by the
Divine power constraining man's obedience. I
only say that he becomes naturally redeemed, so
that his nature will no longer prove an obstacle,

but only a help to his spiritual progress.  In the
past history of the world, men have been regenerated only by the Lord's power, *working in opposition to their nature;* and very few, consequently,
have been regenerated.  And the heights to which
the regenerate have attained, no doubt have been
comparatively below those to which they will attain in the future by the reconciliation of the
natural principle.

No one has ever been regenerated in the least
degree, save by virtue of the Lord's life in man,
that is to say, by virtue of the essential humanity
of God, and the ultimate complete redemption
which that humanity implies for our nature.
Had this redemption been impracticable, there
could have been no basis for the regeneration
even of the very few who have been regenerated
in the past.  (See what Swedenborg says, and
especially promises, without however performing,
in his *Coronis*, 21, particularly VII.)  Regeneration is only a type of the Lord's glorification, or
of the Divine NATURAL man, and has no significance underived from that.   There is no angel in
any heaven at this moment, who enjoys his appropriate bliss by any other tenure than the truth
of the Divine NATURAL humanity, or of that perfect union between the Divine and the human,
the infinite and finite, which takes place in the
*spontaneous* depths of the soul, and which is now
overturning all things in heaven and earth, in

order that it may shine forth in unclouded splendour.

The spontaneous life of man is as yet obscured under vicious institutions. It is the life which science alone inaugurates. It most strictly *presupposes* (and must therefore never be confounded with) 1. The instinctual or animal life, the life of infancy in man, in which the passions dominate the intellect: nor, 2. With the voluntary or moral life, the period of adolescence in man, in which the reason learns to transcend the passions, and rules them by truths. It supervenes only when the course of these things has been run, and man weary of being his own Providence, filially submits himself to the Divine. Instinct is born of the passions ruling the intellect. Will is born of the intellect ruling the passions. The spontaneity is born of a perfect marriage or union between the two, causing all conflict to disappear.

Yours truly,

———.

## LETTER IV.

Paris, Oct. 6th, 1856.

MY DEAR W.

WE are informed that Christ's flesh saw no corruption, and we know that He told His disciples, after His resurrection, to handle His body and make sure that He was no spirit, but an actual flesh and blood man, just as they had always known Him. Swedenborg says that the difference between this glorification of the Lord and ordinary regeneration, or, what is the same thing, between the Divine *natural* man and the angel, is the exact difference between being and seeming, between substance and shadow, between reality and semblance or appearance. Christ's experience was peculiar, or different from the angel's, in this—that He glorified *Himself*, or united His natural selfhood with the Infinite Divine Love. The angel in regeneration does not glorify himself, but the Lord: that is to say, his natural selfhood becomes laid aside, and a new one divinely substituted in its place, so that instead of uniting himself naturally with God, as Christ did, the angel perpetually remits or rejects his natural

selfhood to the devil, and receives from the Lord
an absolutely new self thereupon.  Christ over-
comes hell by His own proper power or manhood,
whereas the angel would be incontinently overcome
of it, if he were not sedulously preserved by the
Divine power, vanquishing his incessant natural
gravitation towards hell.  In short, Swedenborg
affirms that he found no angel in any heaven,
however elevated, who was not in himself, or
intrinsically, of a very shabby pattern, and who
did not, therefore, cordially refer all his goodness
and wisdom to the Lord ; and he sets it down as
the fundamental principle of their intelligence,
that they ascribe all their good to the Lord, and
all evil to the devil.   No matter what heights of
manly virtue the angel may have reached, no
matter what depths of Divine peace and contentment he may have sounded, Swedenborg invariably
reports that in himself, or intrinsically, he is
replete with every selfish and worldly lust, being
in fact utterly undistinguishable from the lowest
devil.  Was ever testimony so loyal as this?  Was
ever honest heart or seeing eye so unseduced
before by the most specious shows of things?   I
confess the wonder to me is endless.   What other
man in that rotten and degraded generation was
capable of such devotion to humanity?  Of which
one of his contemporaries could you allege, that
being admitted to the most lustrous company in
the universe, being associated with men of an

incomparable worth, and women of an ineffable
loveliness, and seeing on every hand the noblest
manners and behaviour, instinct with freedom, and
flanked by every resource of boundless wealth and
power, he would never for an instant lose his
balance, or duck his servile head in homage, but
stedfastly maintain his invincible faith in the great
truth of human equality?   George Washington is
doubtless an unblemished name to all the extent
of his commerce with the world; but how puny
that commerce was, compared with this grand
interior commerce of the soul; and how juvenile
and rustic his virtue seems beside the profound,
serene, unconscious humanity of this despised old
soldier of truth!   To gaze undazzled upon the
solar splendours of heaven, to gaze undismayed
upon the sombre abysses of hell, to preserve one's
self-respect, or one's fidelity to the Divine name,
unbribed by the subtlest attractions of the one
sphere, and unchilled by the nakedest horrors of
the other, implies a heroism of soul which in no
wise belongs to the old Church, even in its highest
sanctities, and which leaves the old State, even as
to its most renowned illustrations, absolutely out
of sight.

But I am digressing.  Our business, thank God,
is not with Swedenborg or General Washington,
or any other person, however eminent, but solely
with the difference between the Lord and the
angel, or the Divine *natural* and the Divine *celestial*

man; in short, between the fully glorified and the merely regenerate aspect of human nature. The difference is indeed enormous, and obvious, moreover to the least reflection. The Divine *celestial* man is comparatively impotent or imperfect, because the angel in whom he is manifested, is not positively good, is not good in himself, but good only by undergoing a process of defecation, that is, by eliminating evil, or precipitating the hells. The Divine *celestial* humanity, or the Lord as embodied in the angel, has only an imperfect ability to subject evil to His obedience; He has power to cast it out, to separate it from good, but not completely to subjugate it to Himself. This latter power inheres only in the Divine *natural* humanity. Thus the hells are so much waste or refuse human force, so much Divine virtue turned to mere excrement, by the inability of the natural man perfectly to unite himself with God. They stand for so much life as the angel fails to appropriate, in consequence of his defective natural organization. Naturally, or *in se*, every man of woman born (and angels boast no sweeter birth than this) is inveterately prone to evil; hence unless he were divinely extricated from his native tendencies, he would never become an angel. This extrication is accomplished by means of the separation of hell from heaven. Hence I repeat that the hells, as Swedenborg pronounces them, are in every variety excrementitious; they are the

waste, the sloughing-off of the human mind in the
progress of its conjunction with God.

Take an illustration from nature. The law of
health for the natural body is, that there should
be a fixed ratio between its supply and its waste,
that there should be as nearly as possible an exact
balance between the two processes of nutrition
and consumption. We daily perform a certain
amount of alimentation or physical regeneration;
and in order that our human force should not be
swallowed up of mere vegetation, it is necessary
that we undergo a proportionate amount of physical
decay or degeneration. Now this physical law is
only the symbol or image of a great spiritual truth,
is merely the ultimate expression, or fixation, of
what is transacting in the higher region of the
soul. Because on the one hand man *is absolutely
void of life in himself,* his life being in truth
incessantly derived to him from God; and because
on the other hand he *feels* that he is life in himself, or underived life; it is evident either that he
must remain eternally disjoined with God by pride,
and all the other evils engendered from it; or else
that some adjustment be operated between his feeling and the absolute truth of things. Now this
adjustment, which we are all familiar with under
the name of regeneration, necessarily involves
in man a certain outflow of merely natural pride
and turbulence, in other words, a certain decease
to his merely natural self; and a proportionate in-

flow of Divine love and tranquillity, in other words, a certain resuscitation to life in God. I say "necessarily involves," because if the *feeling* which man has of his own independence or absoluteness, should govern his reason, or remain uncontrolled by any interior light of truth, he would obviously never rise out of the animal into the human form of life.

Now the spiritual law which I have just recounted is a *universal* law; it is a law which is inherent in the constitution of creation, and as creation is not diverse, but strictly *uni*-versal, so the law I have mentioned is a universal law. Accordingly when we regard the spiritual universe, which is the universe of the human mind, we shall find it discriminated into two opposing spheres, one *celestial*, answering to the influent Divine life in the universal soul of man, the other *infernal*, answering to the outflowing natural life. In other words, every man is internally either angel or devil, and this by the irreversible necessity of creation, so that if we discard the truth of the Divine *natural* humanity, or cling to the perpetual *regime* of the Divine *celestial* principle merely, hell is seen to be as fixed a feature of the spiritual world as heaven itself, every angel existing, as Swedenborg shews, only by virtue of some diabolic antipodes.

But my letter is growing too long, and I must hasten to a close.

From all that has gone before, we see very

plainly that the angel is only a purified or regenerate natural man, and by no means the fully
glorified or Divine natural man, and we need not
wonder, therefore, that the heaven of heavens
should be unclean in the Divine sight. Indeed,
if the Divine Wisdom could effect nothing beyond
the angelic form in creation, the time must assuredly come when the devil would reign supreme
in the universe, for self-love is an infinitely more
potent principle of action than benevolence. But
the Lord, or Divine *natural* man, was always the
vital secret of creation, was always the true power
of God and the true wisdom of God, the Word of
God, by whom all things both in heaven and earth
were made, and without whom was not any thing
made that has been made. In him is life, and
this life is the only true light of men—the light
which lighteth every man that comes into the
world.

Yours truly,

————.

# LETTER V.

*Paris, Oct. 10th,* 1856.

My dear W.

ALL the literal incidents of the birth, life, death, and resurrection of Christ are an eternal REVELATION of the Divine ways to man. These literal incidents all involve the elements of *time, space,* and *person,* or fall within the realm of nature; and hence they do not themselves constitute the substantial verity of the spiritual creation, but only the perfect natural image, form, or manifestation of that verity. The substantial verity of the spiritual creation is, that God alone is life, and that He gives life to man. But this being a spiritual truth, can only be discerned by the reason of man, and not by his senses, under penalty of defeating the entire possibility of creation. Did we *sensibly* perceive God to be the sole life of the universe : were this truth no less a dictate of *feeling* than of reason, we should be most unhappy. For as in that case we should not *feel* life to be in ourselves, of course we should fail to appropriate it, or make

it our own, and consequently should fail to
realise that selfhood, or *proprium*, which is
the condition of all our bliss, because it is the
source of all the characteristic activity that
separates man from the brute. We should sit
like stocks and stones, leaving Him who *obviously*
was life, to the exclusive appropriation and enjoy-
ment of it. But happily sense or feeling is at
variance with reason in this matter, telling us with
the force of an unsuspected instinct that life lies
wholly in ourselves; and hence it leaves the real
and benignant truth of the case to the exclusive
discernment of reason.

Now reason is dependent for its illumination
upon experience, and consequently it cannot discern that God alone is life and the Giver of life
to man, save in so far as it becomes *experimentally
instructed* on that point. The reason does not live
by abstract truths, but real ones. There is no
such thing in God's universe as an abstraction: it
is a universe exclusively of realities. It is not
true abstractly, or apart from fact, that God is life
and gives life to man; but *really* true, or true to
fact, *true, that is, to the experience of the creature;*
and so far as it is untrue to that experience, it is
manifestly not true at all. Man's rational development demands, then, a theatre of experience,
by means of which he may become built up and
established in the truth. For the truth being that
God alone is life and the Giver of life to man, it

is evident that human reason must strictly ignore it, until it become experimentally demonstrated. I say so much is evident, because no one attributes to man the faculty of intuition, which is the power of growing wise without experience. Had we this power, then indeed we might know the divinest truths without the preliminary discipline of experience. But to say nothing of the utter worthlessness of such knowledge to us, were it actually possible, it is abundantly certain that our reason includes no such faculty: and this being the case, I repeat that reason must obviously remain blind to the great spiritual truth that God alone is Life, and the universal Giver of life, until such time as it shall have become experimentally taught that man is without life in himself, and consequently dependent upon God for it.

This distinctively experimental realm, this needful preliminary sphere of human experience, is the world of *Nature*. The natural world is not the real world any more than my body, or apparent self, is my real self: it is only the seminary or seed-place of that world, just as my body is only the seminary or seed-place of my soul. The natural world bears to the spiritual or real world precisely the same relation which the body bears to the soul, the shell to the kernel, the effect to the cause, namely: *the relation of an image to its projecting substance*. As the body is but an image of its spiritual substance, the soul: as the shell is

c

•

but an image of its material substance, the kernel: as the effect is but an image of its logical substance, the cause: so the natural world, or world of sense, is nothing more than an image, form, or reflection of the spiritual world. It always remains the world of form or imagery, in contradistinction to the world of substance, which, being made up exclusively of affection and of thought derived from that affection, is of necessity spiritual and invisible. Nature is, in short, but the stupendous mirror of superior or spiritual modes of being, and it is therefore idle to demand of her any authoritative or final Divine information. Her function is most rigidly propœdeutical or disciplinary, and it would be quite as absurd to expect any original Divine information from her, as it would be to expect the *Edinburgh Encyclopædia* to give you wit, or your looking-glass to give you beauty. Books pre-suppose wit in the reader, as the looking-glass pre-supposes all the beauty it reflects. JUST SO THE NATURAL WORLD, *which is the world of appearances or phenomena,* PRE-SUPPOSES THE SPIRITUAL WORLD, *which is that of substance or reality,* AND IS UTTERLY UNINTELLIGIBLE WITHOUT SOME LIGHT OR "REVELATION" THENCE DERIVED.

Let us clearly understand, then, that the natural world, or world of sense, is never, and in the nature of things cannot be, that ineffable real world which pertains to the soul alone, and whose fundamental truth is that God is Life and the universal Giver of life. Let us distinctly remember, on the contrary, that it is a purely ancillary world to that; that it is in the strictest truth only the handmaid or lieutenant of that better world. The natural world is but a preparatory gymnastic for the soul, a purely experimental realm of life, a marvellous laboratory, at best, in which man becomes acquainted first with what he is in himself, in order that he may worthily estimate what he shall become by the Divine conjunction. This is the exact way it serves the superior or real world, by breeding in us the experimental conviction that we are absolutely devoid of life in ourselves, in spite of every fallacious appearance to the contrary, and hence disposing us to the grateful reception of life from God.

Yours truly,

———.

# LETTER VI.

*Paris, Oct. 20th,* 1856.

MY DEAR W.,

I SAID in my last that the natural world, being the world merely of phenomena or appearance, is strictly unintelligible without some light or revelation derived from the spiritual and real world. The reason of this is not very far to seek. Every lover of Truth, every one who is able to discriminate between Truth and fact, will not merely admit, but will loyally insist that her infinite and spotless form transcends the sphere of sense, or discloses itself only to the eye of reason. All spiritual existence indeed, all those higher truths which affect the soul's life, truths of the Divine existence and operation, of the Divine Creation and Providence, exceed the capacity of sense, and report themselves only to a higher and interior consciousness. To suppose me, for example, *sensibly* cognizant of an exclusively Divine operation such as my creation is: to suppose me not merely *believing* that God is my life, or rationally assenting to that truth, but also actually *knowing* it, or sensibly perceiving it, is virtually to make my faculty of knowledge antedate my being. It is to suppose me essentially on a level with my Creator, or intuitively cognizant of His proceedings : than which of course nothing can be more contradictory. Man therefore can never be *sensibly*, but only rationally, cognizant of his creation. In fact, as we have already seen, his senses, so far as their testimony goes, must rigidly avouch his absoluteness under penalty of utterly vacating his personality. For if we really felt, *i. e.* sensibly discerned, God alone to be life, and thus felt, or sensibly discerned our intimate and incessant dependence upon Him for it, we should fail, as I said before, to appropriate or make it our own, and hence should forfeit all that selfhood or personality which is now contingent upon such appropriation.

If now these things be so : if we have no sensible knowledge of spiritual existence : if from the very necessity of the case, our senses give us no authentic information either as to our origin or our destiny : then clearly all our rational convictions upon the subject must be strictly contingent upon some supernatural illumination, upon what men have called REVELATION, in order to distinguish it from mere information. Revelation is commonly conceived of as if it were only information of a higher grade. Thus Swedenborg is sometimes spoken of by his putative followers as having made a " revelation " of spiritual laws

to us; and the American *spūks* and table-turners
are said by some of their disciples to make *revela-
tions* even in advance of his. But this is an
absurd use of the word. What is meant to be
said in either case is, that the persons in question
give us *information* of a certain character, and the
character being unusual, it is supposed to be entitled on that account to the more dignified name
of *revelation*. But revelation is not a more elevated information, because, strictly speaking, it is
not information at all. Information always means
*imported* knowledge, knowledge which is not involved in our consciousness, but which comes up
to the soul from the senses, or is derived *ab extra*.
Revelation, on the contrary, means *exported*
knowledge, knowledge which belongs wholly to
the sphere of consciousness, or comes down to the
senses from the soul, thus *ab intra*. Whatsoever
the broad face of Nature, or the testimony of
friends, or the lettered page, brings to our eyes,
and ears, and other senses, is information. It
always bears the stamp of direct or immediate
knowledge, being addressed by the sense to the
reason or soul; whereas revelation is always reflected or mediate knowledge, being addressed
*through the consciousness* to the sense as to a
mirror. Information postulates Nature as absolute
or fixed. Revelation, on the other hand, turns it
into a fluent mirror of superior existence. Information is *subjective* knowledge; that is to say, it

embraces whatsoever lies *below* myself, whatsoever is contained in the sphere of sense, and excluded from that of consciousness. Revelation, on the other hand, is *objective* knowledge: that is to say, it embraces all that is *above* myself, whatsoever is included in the sphere of consciousness, and excluded from that of sense. My senses tell me what is below myself: and this is the realm of information. My consciousness tells me what is above myself: and this is the realm exclusively of all Divine revelation. That I can only discern my proper face as it is reflected in a looking-glass is a most strict effect from a spiritual cause; that is, it is a strict correspondence or symbol of the spiritual truth that the soul or selfhood is incessantly derived to us from God, and hence is capable not of an absolute or sensible realization, but only of a *conscious* one.

It is indeed manifest from all that has gone before, that only this revealed or *mirrored* knowledge of spiritual substance is possible to us. God, or perfect Love and perfect Wisdom, is the sole and universal spiritual substance; is what alone gives being and gives form to all things; in other words, is sole Creator and Maker of the universe. But we cannot know God intuitively, for in that case we should require to *be* God: we can only know Him experimentally, that is in so far as we become subjectively conscious of being animated by perfect Love and perfect Wisdom:

or in Swedenborg's formula, in so far as we become "SUBJECTS in which His Divine may be as
in Himself, consequently in which it may dwell and
remain." In short, we know God only through
the Christ, only through the Divine becoming
Immanuel, God-with-us, partaker and glorifier of
our nature, glorifying it to its actual flesh and
bones, and hence as solicitous to give us bodily
health and blessing, as health and blessing of
soul.

Clearly, then, it would not be one whit less
contradictory to postulate for us direct or intuitive
knowledge of spiritual substance, than it would be
to postulate an acquaintance with our own persons, independently of the looking-glass. In the
very nature of things, all our knowledge of the
spiritual world must be reflected, or symbolic
knowledge must come to us precisely in the way
that the knowledge of our own face comes to us,
that is, *through the mirror* WHICH NATURE HER-
SELF IS, *and which therefore she indefatigably
holds out to us*. Of course, so long as we foolishly
hold nature to be destitute of this higher significance : so long as we hold the natural world to be
essentially magisterial instead of ministerial ; to
be, not the mere mirror of the real and substantial world, but that real and substantial world
itself : we shall remain utterly devoid of spiritual
insight, and continue to be the righteous prey of
the dull and lurid despotisms which, under the

names of Church and State, drink up God's boun-
teous life in the soul of man, and turn what should
be His blossoming and fruitful earth, into desola-
tions fit only for the owl, and the fox, and the
bittern to inhabit.

Yours truly,

——.

# LETTER VII.

*Paris, Nov. 1st,* 1856.

My DEAR W.,

You know it is essential to the perfection of a mirror that it should be achromatic or colorless, in other words, *that it should not illuminate the object projected on it.* If the mirror possess any active qualities, any character or significance apart from the function it performs, it will not be sufficiently passive to the impression made upon it; it will of necessity add something to, or take something from, the image sought to be impressed on it. If it possess substantive life or character,—a life or character of its own, and independent of its properties as a mirror, it will of course reflect the image only in so far as the image is congruous with itself; that is, it will not reflect but absorb the image we seek to impress on it. And its value consequently as a mirror will be destroyed. Thus a mirror in order to be perfect must be merely a mirror, *deriving all its force or character from the function it fulfils,* and having no significance independently of that function. If it have any independent significance, any light or life of its own, and apart from its function, it will of course color the image impressed on it, and so far forth vitiate it. A mirror therefore, I repeat, should be as near as possible perfectly achromatic or colorless, imparting no illumination to the image impressed on it, but simply reflecting it as it stands illumined by some superior light.

Now, as we have seen, Nature is the perfect mirror of the Divine image in creation. But inasmuch as every true mirror is achromatic, or incapable of illuminating the image it reflects, we do not ask of Nature *to supply the light* also by means of which the Divine image becomes revealed in her. If we do this; if we allow Nature to dominate the image impressed on her; to color it by her own light; we shall utterly fail of any valid result. A light above nature must evoke the desired image—a light which is not only unaffected by Nature's perturbations and diffractions, but actually reduces them by its stedfast and commanding glow to a perfect order or unity. This light is REVELATION. Revelation is the lamp which darkens the light of sun, moon, and stars, or turns them into the obedient vassals of the soul, by shewing them to be merely a passive mirror of the Divine image in humanity. But here let us always remember that revelation, like man whom it serves, *and whom it could not otherwise serve,* claims both a body and a soul, both a letter and a

spirit. Its letter is fixed or finite: its spirit is
eternal and infinite, that is, is utterly devoid of
relation either to time or space. In its literal or
bodily form, Revelation serves only to dislodge
the mind from Nature's bondage, to break the
shackles of routine and tradition, and so prepare
a ground for its spiritual advent and recognition.
It is a seed—an egg—which being deposited in the
earth of the natural mind will become at length
divinely quickened, and bring forth fruit of a
ravishing spiritual savour. In short, viewed literally, Revelation has no end but to call out or
edify a *typical* church, to separate a purely *formal*
nation to God, or to write His yet unimagined
name and quality on a certain *representative* people, in such large ceremonial characters as shall
vividly image to the cultivated sense, to the eye of
reason, the great spiritual truth of creation, and
so give eternal anchorage or embodiment to whatsoever the heart conceives of good and the intellect
of true in the relation of universal man to God.
For in its spiritual or substantial scope, Revelation
means neither more nor less than the life of God
in the *universal* soul of man, and hence is adequately appreciable only to the most advanced
scientific culture, only to the perfected reason, of
the race.

Let us then be conscientiously careful, my
friend, not to suffer the letter of Revelation to
dominate its spirit. Revelation is indeed a light

to illuminate the Divine image impressed upon
the natural creation, but then as every light supposes an eye to which alone it is addressed, we
must be extremely careful not to darken her lustre
by applying to it the merely natural eye—the eye
of sense. Revelation is addressed only to the
spiritual eye, the eye of culture, to the scientific
reason of man, and we utterly defeat it therefore if
we attempt to comprehend or interpret it naturally.
Science alone—the most advanced science, the
science of those spiritual laws which lie behind
nature, and pronounce her a realm of mere appearances or phenomena—authenticates Revelation. That is to say, science in this enlarged form
corrects our sensible impressions of nature, corrects those purely sensuous judgments of the mind
which break Nature up into endless discords and
diversities, and thus moulds her into perfect order
or harmony. And when we shall see Nature's
order and harmony, we shall assuredly see her
reflecting in every glorified lineament and feature,
the LORD, or that Divine NATURAL man who is
the Alpha and Omega, the Beginning and the
Ending of creation, who is, and who was, and who
is to come, the All-powerful.

Exactly here in fact lies the imbecility of all
our existing theologies and philosophies. The
reason why they all fail to discern the Divine
image stamped upon Nature, is, that they are not
purely rational or scientific, but are all more or

less blinded by the light of Nature. Instead of
viewing Nature simply as a mirror of the Divine
creation, and therefore seeking the image she
reflects by means of a light wholly superior to,
and indeed exhaustive of, her own, our fashionable
theologians and philosophers take her for the veritable creation itself, and listen to her muddled
oracles as though they were the indisputable wisdom of God. The fact is, that our leading authorities in Church and State have not even begun
as yet to extricate us from Nature's initial fallacy
or blunder, which is her affirmation of the absoluteness of the selfhood in man. Nature tells me
by means of my senses, that I am what I am absolutely, that is, without reference to other existence: it says that all my good and truth, all my
affection and thought, originate in *myself*, and
consequently when either good or evil prepon-
derates in that affection and thought, it bids me
appropriate the merit or demerit thereof to my
*self*; thus making me at one moment the victim
of the most baneful moral elation, and the next of
an almost equally baneful moral depression. Self-
conceit and self-reproach, pride and penitence;
these make up the fever and the chill into which
that great intermittent, which we call our religious
experience, ordinarily resolves itself. In short,
Nature taking advantage of the selfhood (which
is God's ceaseless communication to me, which
constitutes, in fact, His inalienable presence in

me, the true Shekinah or holy of holies), to affirm
my absoluteness or sunder my solidarity with all
my kind, drives me to an eager eating of the tree
of knowledge of good and evil, so shutting me up
to an alternate moral inflation and collapse, and
in both cases alike disqualifying me to eat of the
tree of Life.

The service which Swedenborg has done the
rational or scientific mind, by the light he has
cast upon this great truth of human solidarity, is
incalculable.  He proves to us by a faithful exposition of spiritual laws, which are the laws of creation, that no individual is independent of any
other, and that there is consequently no such
thing as individual approbation or individual condemnation, in the Divine mind.  He shews us
that, since the world has stood, *no man has been
chargeable before God with either his moral good
or evil*, because neither the one nor the other originates in the man himself, but are both alike an
influence from other beings with whom he is spiritually associated.  He shews me that all the
good I feel in my affections, and all the truth I
realize in my intellect, are an indubitable influence from heaven; and all my evil and falsity a
like influence from hell.  Both good and evil,
truth and falsity, flow in to the natural mind un-
impeded, because the natural mind being the
*common* mind of the race, is the sole basis or continent of all its spiritual good and evil, and is

indeed vivified solely by giving these things unity.
But this being the case, if I proceed thereupon to
*appropriate to myself* this influent natural good, or
this influent natural evil : if for example when I
have done good to my neighbour, I look up to
God with a sense of self-complacency, feeling that
He loves me now *more* than He did before ; and
when I have done evil to my neighbour, I look up
to God with a sense of ill-desert, feeling that He
now loves me *less* than He did before : I then
exclude myself from the tree of Life, the life of
the Lord, or the Divine NATURAL humanity, and
shut myself up in eternal death which is stupidity
—the stupidity that grows out of a *cultivated*
self-satisfaction.    From the beginning mankind
has known no other curse than this, *"eating of
the tree of knowledge of good and evil ;"* and yet
it is the curse which all our ecclesiastical and
political doctors, backed by all our sentimental
and professedly infidel scribes,* are assiduously

---

* There are no truer friends of the infantile Church and State
than Mr. Francis Newman and Mr. Holyoake : and if the Catholic
Church were not providentially doomed and done for, or what is
the same thing hopelessly besotted as to her own interests by her
own selfishness, she would instantly remit John Henry Newman
to the back settlements, and exalt Francis and the Secularists into
his place.    For these men are the real *ames damnées* of Moralism,
heartily despising and deriding Christ for every utterance and every
act which expressed a Diviner life in him (and if in him actually, so
of course by implication potentially in all men his brethren), than
the mere life of conscience or will.    And this is the precise temper

busy in fastening upon us.  One can hardly ex-
aggerate the zeal they display in this disreputable
calling ; but one can easily anticipate the fierce-
ness of the reaction which, under the providential
illumination of the scientific conscience, they are
preparing for themselves ; and which will leave no
vestige of their futile labours surviving.

                              Yours truly,

                              ———.

of mind which vivifies our present ecclesiastical and political corruption : a disposition to aggrandise and intensify the *moral* life of
man, to inflame that sentiment of personal difference and distinction among men, which litters all our existing contrasts of good
and evil, rich and poor, wise and simple, proud and grovelling,
and so immortalizes the reign of hell on earth.   I am perfectly persuaded that Mr. Newman and the Secularists are as unconscious of
any desire to oppress mankind as I am, and I do willing justice to
the benignity which animates much of their writing : but when the
best-intentioned persons persist in prescribing for diseases which
they are obstinately content to know nothing about, better instructed
people grow tired at length of acknowledging their infatuated good
intentions, and reasonably vociferate for some slight increase of
understanding as well.

# LETTER VIII.

*Paris, Nov. 7th,* 1856.

MY DEAR W.

It is a well-known law of optics that every image of a natural object impressed upon our retina, takes an *inverted* form. For example, I am looking at a horse passing my window. Now if you could look into my eyes, you would perceive that the image projected from this horse upon their retina, was upside down. Natural philosophers, as they are called (by those who conceive that Nature furnishes philosophy as well as fact) are very much like that incoherent gentleman who tried to lift himself from a lower to an upper story, by tugging at his waistband, and accordingly always insist upon making Nature explain herself, or give an account of her own processes; than which nothing can be more unphilosophical. For Nature is a sphere of *effects* exclusively, and utterly banishes *cause* therefore beyond her borders. To ask Nature to give you an insight into creation, or to shed any light upon her own causes, is in reality not a bit less absurd

than it would be to ask your coat and pantaloons
to give you an insight into humanity, or explicate
their own genesis in the moral and meteorological
necessities of their wearer. Nevertheless our
natural philosophers view things differently, and
have long been trying to account on optical principles exclusively, for our seeing things upright,
*in spite of the invariably inverted form they take
upon the retina.* They say they find it easy enough
to account on natural principles for the inverted
form itself; for this is a purely optical fact growing
out of the relations of the eye to light, and needs
only the knowledge of those relations to explain
it. What they call accounting for it in short, is
only stating its constitution.

But *our seeing things upright* in spite of the
inverted image they take upon the retina! In
fine, the bare fact of SIGHT at all! Ah! this is
by no means a mere optical fact, and refuses to be
explained upon purely natural principles. It is a
fact utterly transcending the realm of optics, and
your most ingenious Herschels and Aragos are not
a whit better qualified to pronounce upon it than
you or I, or any other simply honest person. For
sight, considered as active, is a fact of Life exclusively, of life manifesting itself no doubt by an
organized medium, but entirely unidentical with
such medium. In short, sight is a purely SUB-
JECTIVE experience, the experience of a living subject, and refuses to become intelligible save by re-

ference to that Supreme Life or Being to which we are all variously but equally subject. To understand the phenomena of sight accordingly, to perceive the reason why, for example, we see all things in nature upright, when their images are reflected upon the retina in an inverted form, we must not merely know the laws of optics, but we must above all things know the laws of Life, life universal and particular. And those laws as we have already seen, demand for their elucidation a light above that of the sun, demand in fact a spiritual Revelation such as Christianity purports to be.

Let me attempt to approach this great theme, then, in a way which will not too violently shock your prejudices, by seeking to explain the phenomena of natural vision, or to shew why we invariably see horses and cattle, houses and trees, upright, when their images are always reflected upon the retina upside down.

The summary explanation of all natural experiences, and this among the rest, is, that Nature is but an *experimental* world; in other words, that from her lowest pebble up to her perfected form which is the human body, she is but a *mirror* of the soul, or true creation : and it is never the function of a mirror to reflect that which inwardly or really or consciously *is*, but only that which outwardly or actually or unconsciously *appears*. In other words the mirror never reflects being as

it exists to itself, or consciously, but only as it
exists to others, or phenomenally. Thus it never
imparts intelligence or wisdom to us, but only
fact or appearance which are the servants of wisdom. If for example I should visit my glass every
morning for instruction as well as information, for
wisdom as well as knowledge : if I should go there
not merely to ask how I *appear* to others than
myself, but also to ask how I really exist to myself : I should instantly find every dictate of my
consciousness belied. I should be sure to put the
patch which belonged to my right cheek upon the
left, and give my left whisker the trimming which
every interest of equilibrium demanded only for
the right. For the mirror invariably tells me
that my right hand is my left, and my left hand
my right, so that if I were to obey its instruction
for the real truth of the case, instead of depending exclusively upon my own natural consciousness,
I should soon exhibit as insane a picture personally,
or with reference to the interests of my body, as
*he* does spiritually or with reference to the interests
of the soul, who follows the teaching of Nature
in that regard, without reference to the commanding light of Revelation. In short our mirrors
never disclose the veritable being of things, but
only the form or appearance which that being puts
on to other eyes than its own. They do not give
us the perfect Truth itself, but only the mask, the
appearance, the semblance which that truth wears

to an imperfect or finite intelligence, as to the
bodily eye for example: and we seize the essential
or rational Truth in every case, by exactly reversing this mirrored or reflected semblance of it.

Now nature being a mirror of the soul or spiritual creation, and nothing but a mirror, we must
of course insist upon her renouncing all higher
pretensions, and observing strictly every exigency
of her own character. That is to say she must
reflect the soul or spiritual world, not as it is in
itself or really, not as it exists to its own consciousness, but only as it is phenomenally, or as
it exists to a more limited intelligence than itself,
say the bodily eye. In other words, we must
expect to see the natural consciousness exactly
reversing the spiritual one. Thus what the spiritual affection pronounces good, the natural
affection must pronounce evil: what to the spiritual understanding is truth must be to the
natural falsity: what is light to the spiritual eye
must be darkness to the natural eye: what to the
former is right must be left to the latter: what is
head to the one must be heels to the other: and
so forth. By natural light therefore, *the light
which the mirror herself supplies,* it is no wonder
that all things within her framework appear upright and orderly and beautiful; just as in the
looking glass that which is really or consciously
my left hand is made to *appear* my right: whilst
in reality or to the spiritual consciousness, they

are the exact reverse of upright and orderly and
beautiful, as we see by the inverted forms they
assume when they are reflected towards the soul,
whose nearest outpost is the retina, and other
apparatus constituting the needful basis of the
varied life of sense.

It is accordingly *not at all remarkable* that we
see things upright, whose image upon the retina
is inverted, because natural light, by which we
see the things in question, is in itself but an inversion or correspondence, and by no means an
extension, of spiritual light.  Those who make a
marvel of this experience, undoubtedly hold that
*we see the image* of the natural object on the retina,
instead of the natural object itself.  But this is
simply absurd or contradictory.  It would indeed
be truly marvellous, if while actually *seeing* this
inverted image we yet saw the object upright.
But this is impossible.  In that case it would not
be the eye which sees, but the brain.  For we see
not what lies within the eye, but what lies without
it: and the image in question falls not upon the
eye, but exclusively upon the brain through its
extension into the retina.

But some one will ask, Do we not see at least
*by means* of this inverted image?  Do we not see
by virtue of a reflection of the natural world on
the retina?  This question puts the cart before
the horse, but I can manage to satisfy it.  It is a
universal truth that the natural world is altogether

vivified from the spiritual one, and it is also true
that this vivification takes place through certain
media, which we call the senses. Thus we see,
we hear, we smell, we taste, we touch, which are
all experiences of natural life, by virtue of a spiritual influx into the retina, the tympanum, the ol-
factory and gustatory nerves, and the skin. But
this influx does not traverse these various media,
or pass through them : on the contrary, it is
always arrested there by the exact contrariety or
inversion which it encounters at the hands of Nature; and it is this very arrestation which becomes
the basis of our natural subjectivity, or makes our
natural experience possible. Thus the inverted
natural image on the retina is nothing more nor
less than a reverberation or *contre-coup* made upon
the spiritual sense by an act of natural vision : it
marks the arrest and reflection, or bending back,
of the spiritual world upon itself, when it would
otherwise pass out of its sphere, and dominate the
natural one. It is therefore true to say that we
see by means of this reflex natural image on the
retina, thus far, namely :—that if that reflex image did not take place, it would be because there
was no difference between soul and body, between
spirit and nature, and consequently because we
were not intended to enjoy any natural life. Beyond this, it is absurdly untrue.

I said just now that it was this arrestation of
the soul, or of spiritual influx, at the portals of

sense, which alone allowed us a natural subjectivity. This is obvious enough. For if the soul passed through these nervous media, so subjecting the natural body to itself: if in other words the spiritual consciousness dominated the bodily or natural consciousness, passing into it not courteously and by correspondence, but brutally or in person: why then of course we should have no natural sight or natural consciousness of any sort. We should in that case be the mere slaves and packhorses of the soul, and would soon lose even the bodily form appropriate to humanity: for that erect form postulates an indwelling divinity or freedom even down to its toe-nails. In a word natural experience is never, but in most diseased and beastly conditions, the continuation or reproduction of spiritual experience: it is most strictly, or at its healthiest, a correspondence and inversion of it, just as the inside of a glove is a correspondence and inversion of its outside. We see naturally only by ceasing to see spiritually; and we cease to see spiritually by the very necessity of our natural organization, which makes the eye, the ear, and every other sense a *common* medium for the soul instead of an individual one. Spiritual sight demands an organ which is empowered only from *within* the subject, which derives its potency entirely from the affection and thought, or spiritual character, of such subject. Hence Swedenborg continually saw persons whose interiors were of that human largeness that they *sensibly* communicated with the remotest planets, turning the distance between the earth and Orion into a childish superstition, into a mere scientific pedantry. But natural sight demands an organ most strictly irrespective of the individual character of the subject, because depending exclusively upon his relations to the race, or what he has *in common* with all other men. Thus the angel Gabriel if he were in the flesh, could make no morning call in Andromeda, whatever might be his inward fitness, and would be obliged to drop his dearest friend in Arcturus, simply because his natural organization is not the continuance or extension of his spiritual one, but its decisive contrast and contradiction, invariably pronouncing that *first* which the latter pronounces *last*, and declaring that the highest good and truth and beauty, which the latter declares to be the lowest evil and falsity and deformity.

But I must come a little closer to my subject, and by way of relieving your fixed attention, I will postpone what more I have to say to another letter. Don't grow discouraged; the goal is clearly before us, with heaven's own radiance encircling it, and it will not be long before we grow perfectly familiar with the approaches to it.

Yours truly,

———.

# LETTER IX.

*Paris, Nov. 20th,* 1856.

My dear W.

You know how continually Swedenborg protests against the popular habit of regarding space and time as real existences, and how he denies that any right understanding can be had of creation so long as that habit remains undisturbed. " Do not, he says, I beseech you, confound your ideas with time and space, for in proportion as you do so, you will really understand no Divine work. Creation cannot be explained in an intelligible manner, unless time and space be removed from the thought, but if these are removed it may be so explained. It is manifest from the ideas of the angels which are without space, that in the created universe nothing lives but God-man alone, or the *Lord,* and that nothing moves but by life from Him : thus that in Him we live, move, and are."—*Divine Love and Wisdom,* 51, 155, 285-6, 300-1.

Now what does Swedenborg mean by thus everlastingly warning us against taking our sensible judgments of time and space as absolute, or con-

firming them from the reaso
other words, to say that theologi
phers have an inveterate habit of r
as a commerce or play between an
and an *external object*, and so
*brutifying* life.  For as in every
of object and subject, the object
member of the copula, and th
dient one, so of course I cann
external universe without bein
of the reach of my subjection
prived of my human quality.
lies in finding its object on
plane to that of its own subj
*interiorating* the object to th
quently when philosophy
process and shews me to
subject of nature, it deg
*dehumanizes* me.  We
scientific speech, that th
denborg, of all our ort
lies in their sensuality, t
*exterioration* of the obj
*not-me* to the *me*.

Take for example any
sophers call, a sensible
" Now" say the philosop
two things involved in
and an object, a *me* and
ing subject, the horse

...hood images this creative peculiarity, this Divine perfection, putting the creature so far as he
realizes it into exact correspondence with the
Creator, and making Him a full participant or
object of the Divine infinitude.

But how shall the creature realize this selfhood in any degree, seeing that he is absolutely
destitute of it? It is evident from the bare statement of his creatureship, that he is *intrinsically*
*void of life* or selfhood, that his very nature is
*not to be.* How then shall he surmount this
intrinsic or natural destitution, and so arrive at
selfhood, or conscious existence? This is the
question. On the one hand, it is clear that he
cannot be a creature of God, save in so far as he
possesses selfhood: on the other, it is clear that
he is intrinsically, or by nature destitute of selfhood. And the problem is to reconcile these two
propositions, or to shew this intrinsic natural destitution of life giving place to the amplest and
eternal exuberance of it.

But now does it not irresistibly follow from
these premises, that creation is nothing more and
nothing less than a process of redemption? Does
it not follow, in other words, that all true or
Divine creation rigidly consists in giving the
creature redemption from his own nature, that
is, in endowing him with selfhood? Bethink
yourself. We have just seen that God's true
creature is bound by the necessity of his creature-

E

" For example : take any experience of life, say
some fact of vision, as when I see a horse, a tree,
a sunset, or any thing else in Nature : now to all
the extent of this experience I am merely realizing
a fact of consciousness or of relation : I am not
exerting any sensible power, or putting forth any
latent faculty stored away in my visual organ independently of the surrounding universe. Sight
is never in the eye alone, or *apart from* the things
seen. It is only in the eye as livingly *associated*
or fused with the universe of creation, with all
that the sun shines upon. Take away the horse,
and the tree, and the sunset, with whatsoever
may stand in their place, and you take away my
sight. Though I had all the eyes of Briareus, I
should be more blind than a bat : I should not see
at all. For I have no *absolute* power of sight or
hearing, or smell, or taste, or touch ; that is to
say, in myself considered as unrelated to, or disunited with, the universe of light and sound, etc.,
I have no power of any sort, I am even destitute
of consciousness, and do not exist : but in myself
considered as related to or one with all these uni-
verses, I am full of power. Thus the eye is vivified, not apart from, but only in conjunction with,
the universe of light ; and so of all our other experiences, they are none of them simple facts, but
all are *composite* ones, involving our intensest unity
with Nature, or the universality of the *me*. They
are all facts of *con*-sciousness : that is, they all
imply, that though in reflection or when I listen

to my senses simply, I know myself as limited to this wretched body, yet in life or consciousness, when I am acting and not merely thinking of myself, I know myself only as one with the sensible universe, as lovingly blent or associated with all that my senses contain and embrace.

"It is evident from all this that I do not exist consciously as an independent being, but only as a most dependent one, that is, as an individual form involving strictly universal relations. *I* exist only to consciousness, only in so far as I feel myself in *universal* relations; and I LIVE, that is to say, my existence becomes beautiful and delicious to me, just in so far as these relations are relations of complete accord, furtherance, and obedience. When the eye does not spontaneously melt into its own universe, or the realm of light, but shrinks into its bodily enclosure, it is diseased and ready to die. When the ear does not spontaneously command its own universe, or the realm of sound, but recoils upon itself, it no longer lives but is preparing to die, and is only kept alive in fact by the influx of life into the healthier organs. And so of all our senses, the moment they cease to universalize the soul or *me*, the moment they begin to shut us up to our bodily dimensions, they are diseased and prove a curse instead of a blessing, an avenue of conscious death in place of conscious life.

"Understand then that the soul, the me, the

selfhood is a purely conscious existence, unrecognizable by sense. No doubt my body exists to your eye, quite absolutely and independently of all other bodies; and no doubt that foolish eye may identify my body with my *self* or *me*. But this is a pure fallacy of sense. To the senses I exist only animally, or as a natural body, subject to all the laws of nature, and in this gross, sensual, and culinary form I first come to consciousness no doubt, and present myself to the acquaintance of your bodily eye. But I exist humanly and really only to your interior senses, only to those subtler senses which belong to your spirit, and which recognize me under exclusively spiritual forms, since they pronounce me now *good* and now *evil*, now *noble* and now *mean*, now *wise* and now *silly*, now *amiable* and now *detestable*; which are all qualities of spirit and not of matter. In fine then, the *me*, truly viewed, is altogether a conscious existence, or knows itself only in inseparable unity with the universe; and by restricting it to bodily dimensions, or subjecting it to the laws of space and time, you degrade and stifle it quite as much as you degrade and stifle my body when you incarcerate it within dead walls, and seclude it from the genial light and warmth by which alone it lives. Life, consciousness, always implies association, always implies the fellowship, union, or fusion of two *sensibly* distinct or disunited forms, the specific and gene-

ral, the unitary and universal; just as water
implies the fusion or unity of oxygen and hydrogen
for its own production. To the spiritual intelligence accordingly, it would be no less absurd to
separate the unitary element in consciousness
from the universal one, and call the gasping thing
life, than it would be to the scientific understanding to expel hydrogen from oxygen, and call the
crazy and viewless remainder water. Water is
the perfect fusion or union of oxygen and hydrogen, just as the living *me*, the conscious individuality, is the perfect fusion or union of the unitary
and universal life. No doubt that oxygen and
hydrogen in order to form water, combine in
invariably definite proportions; but this is only
saying in analogous terms, that the fusion or
union which the individual consciousness operates
between the specific and the general, between the
unitary and the universal forms, is most strictly a
*marriage*-fusion or union: that is to say, that the
former or limitary element in consciousness is
always feminine, and the latter or universal element is always masculine, and that the secret of
human destiny lies in allowing the former element
the free preponderance of the latter.

Thus, in effect, Swedenborg explodes the popular conception of consciousness, and shews it
utterly unworthy of the reality. For consciousness disavows the antagonism asserted by sense
between the various forms of Nature, and proves

them indissolubly fused and blent in the unity
and universality of the *me*.  In short, consciousness claims the totality of the sensible universe
as the indispensable realm of the *me*, and consequently finds no faintest glimmer of the *not-me*
within it.  When I listen to sense, which has a
very subtle and insinuating voice, I hear precisely
what the philosophers hear: I hear that the distinctively *human* force in me, the soul, the self,
the *me*, is subject to the natural force, is subject
to my bodily limitations, or the laws of space and
time: thus that I stand in the fixed relation of
*subject* to the cat and the dog, the cockroach and
the louse, and all other forms of universal life;
and all these forms again in the fixed relation of
*object* to me.  But when I grow indignant with
this sensual stuff, and listen to the voice of consciousness instead—to the voice of the soul, the
reason or *true* me—I hear an exactly opposite
doctrine.  For the spiritual reason or consciousness tells me whenever I consult it, not that I am
subject to the natural universe, but that the natural universe is properly subject to me, is in fact
merely the contents of my spiritual subjectivity.
It brings the natural universe, by means of the
senses, within the periphery of the *me*, within the
realm of conscious life: and consequently it utterly eliminates the *not-me* from the finite sphere,
binding me to seek it instead in that of spiritual
substance, the sphere of infinite Love and Wis-

dom. In other words, it identifies the *not-me*
exclusively with God, thus denying me, as *subject*,
any proportionate or befitting *object*, short of the
immaculate Divine perfection. And in so doing,
it manifestly stifles Atheism on the one hand, by
proving God the sole life of the universe; while
on the other hand, it sops up Atheism's younger
and feebler brother, Pantheism, in yet separating
God from that universe by all the breadth of our
spiritual consciousness, by all the amplitude of the
finite *me*.

If you are desirous after a fuller investigation
of the constitution of consciousness, I refer you to
an article in *Putnam's Monthly* (New York, No-
vember 1853), entitled *Works of Sir William
Hamilton*. An earnest attempt is there made to
expose the vulnerable body of our orthodox philosophy, with what success I must leave you to
determine. At present I hear you inquiring,
what, after all, Swedenborg's rectification of our
intellectual methods avails to the right understanding of creation, or to a scientific cosmology;
and this question I at once proceed to answer.

I may answer it briefly by saying, that it avails
thus much: without that rectification creation is
inconceivable, is in fact a dense absurdity; and
the old church is simply right in betaking herself,
as we see her doing, in the person of all her
actually living children, of all those whose intellectual life is not swallowed up in mere routine

and formalism, to the embraces either of Atheism or Pantheism. The hardier and intellectual sort among them will prefer the former terminus as effectually ending the journey, and giving the soul a long *quietus*. The tenderer and affectionate sort will prefer Pantheism, as still keeping up some faint semblance of progress, although that progress be decidedly inhuman, or from the solid back into the liquid, and even the gaseous state. But both sorts alike are the legitimate children of the orthodox church, and do but illustrate the logical dilemma into which her prevalent Naturalism forces all her honest and clear-sighted descendants. Nothing short of this explains the church's hatred of them, and her eager disavowal of intellectual complicity. She evidently feels herself endangered by their inability to keep counsel, and hates them accordingly very much as the convicted culprit hates the treacherous *approver*.

But you do not desire so brief an answer as this, and I had therefore better commit what I have to say to another letter.

Yours truly,

————.

# LETTER X.

MY DEAR W.

I AM now going to discourse to you a little while about the two following propositions: first, that creation is strictly unintelligible on the orthodox hypothesis of personality; and, second, that it becomes strictly intelligible when you substitute an improved or scientific conception of that subject.

The orthodox hypothesis of spiritual existence, or of the *me*, imports that I am quite as absolute or finite with respect to my soul, as I am with respect to my body. It supposes that spiritual existence is equally absolute with physical, and consequently has as little dread of the conscience pronouncing me good or evil, amiable or hateful, and so limiting my spiritual personality, as it has of the senses pronouncing me blond or brown, handsome or ugly, so defining my natural personality. It accepts without any misgiving the insurgent dictation of the senses in this particular,

and looks upon the selfhood, or personal element
in me, as spiritually claiming the same rigid
fixity, the same absolute dimensions as my body.
For example, I steal, commit adultery, or murder.
Conscience tells me that these are evil and abominable deeds : and my self-consciousness, instructed
by the current theology and philosophy, appropriates this evil to my*self*, or pronounces me an
evil man, justly abominable to God and all good
men. There are others, who unlike me, refrain
from all these misdeeds, who in all their domestic
and civic relations strive to fulfil the golden rule,
and do as they would be done by. *Their* self-
consciousness, again instructed by the orthodox
philosophy of the selfhood, affirming its essential
absoluteness, appropriates this good to them, or
pronounces them good men, entitled to expect the
blessing of heaven upon themselves and their pos-
terity.

Manifestly, then, the orthodox notion stultifies
itself. For it presents us two distinctly opposite
beings claiming the creatureship of one and the
same infinite power. It presents us two beings as
vividly contrasted as evening and morning, only,
unlike evening and morning, which both alike
but in successive order melt into perfect day,
these contrasted beings declare themselves absolute or unrelated, and refuse to merge therefore in
any higher and unitary personality.

So sheer a contradiction as orthodoxy here

offers us, forces us of course upon one of two conclusions : either 1. that the good and evil man are not the final subjects of God, but only the intermediary and transient form of that subjectivity; in other words, that the moral life is not the true life of God in the soul of man : or 2. that the orthodox philosophy is wrong in her estimate of these men, they being not absolutely good and evil as she affirms, or good and evil with respect to each other, but only phenomenally so, or with reference to the divergent relation they bear to another and higher life. In short, their good or their evil cannot be attributed to themselves individually, and hence does not characterize them in the Divine sight, but must ascribe itself, all that is good in them, to their common creative source, exclusively, and all that is evil in them, exclusively to their common formative nature.

No sane man can deny moral distinctions. The distinction of good and evil, truth and falsity, among men, is as palpable to the soul, or rational experience, as that of heat and cold, light and darkness, is to the bodily experience. The reason indeed is vivified by those differences, so that if you annul them you evaporate reason itself. But it is only the more clear, therefore, that such vital opposites, if you regard them absolutely or in themselves, and as unaffected towards some third and neutral term, cannot acknowledge the same

creative source. The philosopher may indeed
allege, that he does not mean to say that the evil
man was created evil by the Divine hand, but
that having been originally created good by that
hand, he afterwards became evil of himself. But
who but a determined suicide, leaps into the fire
in order to save himself from the frying-pan?
For to say nothing of the possibility which is here
admitted of any good man extant losing his present status, and becoming converted into an evil
man, one is immediately prompted to demand
where he who was originally created good by the
Divine hand, got the power to defeat that creation, and render himself evil? If the answer be,
that he was created with that power, then as we
can't conceive God giving a power to his creature
which He would not have the creature exercise
or enjoy, you evidently make God a participant
in the downfal of His creature. If the answer
be that the power was derived from the Devil,
the question immediately recurs, and remains in
fact insatiable, who is this Devil that thus mas-
ters God, and converts a Divine performance from
good into evil?

I repeat, then, that we cannot regard the good
and evil man as true creatures of God, save in so
far as we cease to regard them absolutely or in
themselves, and view them exclusively as they
stand related to a third term, which shall have
power to annul or swallow up their intrinsic anta-

gonism and conflict in the breadth of its own majestic unity.   In other words, the moral realm
cannot be regarded as the realm of the Divine
creation, unless you make moral existence to be
purely elementary and subsidiary to an infinitely
superior life.   If we make moral distinctions absolute: if we make them to attach to men not
merely in their own finite estimation, but also in
the Divine estimation: if, for example, we say
that Mummy the murderer, as compared with
any indubitably good man, say Dr. Channing, is
and always will be a bad man, not merely to our
judgment, but also to God's judgment, so as that
God will really love the one man and hate the
other, or at all events experience a conflict of
emotions in reference to them: it is evident that
we instantly destroy the Divine infinitude, or invest Him with a fickle perfection.   And if we
start from any *datum* short of God's immutable
perfection in constructing our cosmology, it is
certain that we must never expect to achieve any
satisfactory scientific result.   A scientific cosmology is totally inconceivable upon any other postulate than that of the complete dependence and
equality of the creature in the creative estimation.
It must hold with Swedenborg and the more instructed angels, that all men are precisely alike in
the Divine sight, good and evil, wise and silly,
strong and weak, rich and poor; and that all their
differences arise not from any absolute root, but

from their various relation to the Lord or Divine
*natural* man, whose true or spiritual advent is now
taking place in all the abounding truths of human
fellowship or unity.   Swedenborg shews us that
the good man, the saint, the angel, the seraph,
name him as you will, is in himself or intrinsically
of exactly the same pattern with the evil man, the
sinner, the devil, the satan; and that all his differential good confesses itself to be of the Divine
operation in him, continually subjugating his intrinsic tendencies.   I once knew a loquacious person
who said : " I can't imagine how any one should
have any distrust of God.   For my part, if I were
once in His presence, I should feel like *cuddling-
up* to Him as instinctively as I would cuddle-up
to the sunshine or the fire in a wintry day."   It
is beautiful to observe how utterly destitute Swedenborg found the angelic mind of all this putrid
sentimentality, this abject *personal* piety.   He
never met with any angel rich enough to patronize Deity, or to imagine that God felt the least
personal affection towards him more than He felt
towards the duskiest denizen of hell.   In short,
he found the angels of an intensely human quality, or saved from lying and theft, adultery and
murder, not by feeling themselves or thinking
themselves any better than other men, the most
infernal; but simply by feeling and thinking
themselves intensely *one* with all other men, even
the most infernal.   In a word, the ground out

of which all angelic manhood springs, is never
Pride, or the sentiment of personal difference
among men, but always Humility, or the sentiment of their complete spiritual unity.

Let us then abandon the orthodox cosmology
as simply incredible, because it makes the *me*
absolute, and thus sows division in God's spiritual
universe: because, in other words, it makes the
spiritual or real world to consist not of one richly
unitary life or soul (which we may name *maximus
homo*, or grand man, in contradistinction to you
and me as *minimi homines*, or least men), but of
as many discordant and disunited souls as nature
presents of discordant and disunited bodies.  This
is what the good book calls " eating of the tree of
knowledge of good and evil," at the instigation
of the serpent.  The serpent symbolizes the sensual principle, and " eating of the tree of knowledge of good and evil," means accordingly a state
of understanding in man instructed only by the
senses.  In the infancy of the natural mind, whether of the race or the individual, we judge only
according to the sensuous appearance of things,
and not according to their rational reality: in
other words, we *appropriate to ourselves* the good
and the evil which are derived to us only from
spiritual association, being utterly ignorant that the
one is strictly an influence from heaven, and the
other an influence from hell.  And thus appropriating these things to ourselves, we are inevitably

filled with self-complacency or self-loathing, just
as the one or the other influence prevails. The
eternal Wisdom in its letter says to man: *Thou
shalt not eat of the tree of knowledge of good and
evil without the surest experience of death:* which
being spiritually interpreted, means: " Cease to
appropriate to yourself either the good or the evil
you know, because neither the one nor the other
belongs to you, both alike being intended for no
other purpose than by their exact equilibrium to
give you selfhood, and so provide a basis for your
subsequent immortal conjunction with all Divine
perfection.   If, therefore, you foolishly appropriate
this merely influent good and evil to yourself, you
will completely misconceive your destiny: you
will so far defeat the Divine benignity towards
you.   Instead of becoming Divinely stript of all
baggage, of all natural impediment, and so qualified to aspire after all Divine perfection, you will
become inflated with pride, feeling yourself as
knowing as God; and hence, instead of filially
following His ways, and finding peace therein,
you will insist upon being your own Providence,
and will thus manage, unless the Divine wisdom
counteract you, to immerse yourself in endless
perplexities and bring up in final despair.  In
truth, by persisting in this insane career, you will
grow so full of inward death, so replete with malignant pride and self-love, that you will compel
the Divine love to defecate you of your own self-

hood, of your proper nature, and endow you with a new one more surely pliant to His great behests."

Swedenborg, in his amazing pictures of the spiritual universe, so dull and unattractive to the thoughtless mind, but so vivid with every charm of colour to the instructed sense, shews us this actual defecation of the natural selfhood, or *proprium*, going on in the separation of hell from heaven; and he proves, against all rational cavil, that the eternal and intimate sweetness of the Divine creation is contingent upon such separation. He never for a moment represents these transactions in the realm of man's spiritual experience, as an arbitrary arrangement, or as being their own end. On the contrary, he conclusively proves that they grow out of the very necessities of creation, being rigidly subservient to the permanent redemption of the human mind from the baseness and stupidity it contracts, in listening to the flattering and fallacious dogmatism of sense. And this brings me to my second point, which is to shew how intelligible creation becomes, when you take an improved and rational view of the constitution of the selfhood. But I will treat this point in another letter.

Yours truly,

————.

# LETTER XI.

*Paris, Dec.* 3, 1856.

My DEAR W.

My last letter went to prove that creation was strictly unintelligible, so long as you made the soul, or the *me*, an essentially finite existence like the body; so long as you identified it with the dimensions of time and space, and so separated it from the unity and universality of Love and Wisdom.

I propose in the present letter to give you the converse aspect of that proposition, or to shew how strictly intelligible creation becomes, when you view the soul, or the *me*, as essentially infinite, or infinite in itself, and finite only by subjection to the individual or natural consciousness, the consciousness instructed by sense. Let us proceed, then, with all possible dispatch.

If, as we have already seen, time and space are not real and substantial existence: if, in truth, they are only the semblance—the appearance—which that existence puts on to a lower range of intelligence, say to the senses: then clearly the

soul can avouch itself a real and substantial existence only in so far as it casts off the subjection of
time and space. But that which is not subject to
space and time is infinite and eternal : and infinite
and eternal are two words employed exclusively
to designate uncreated being. Thus the soul, in
order to avouch itself a real or substantial existence, is bound to be infinite and eternal, is bound
to be uncreated.

Viewed spiritually, then, the soul is uncreated,
is in simple verity, God. How then does it become what we call created, that is, subject to space
and time? For this is what we invariably mean
when we call a thing *created ;* we mean that it is
a finite existence, that it is subject to the conditions of space and time. How then does the
soul, which is essentially uncreated, being the infinite God, become as it were converted into a
creature, become finited in space and time? Swedenborg sheds a flood of light on this inquiry, the
most interesting that can engage the mind of man,
and makes it intelligible to the plainest capacity.
He shews that the soul becomes created, or falls
under the dominion of time and space, not really
or to its own apprehension, but only apparently,
or in accommodation to the exigencies of our selfhood, of our individual consciousness. For example, I have no absolute or underived selfhood, but
only a reflected or derived one, such an one, in
fact, as I derive from my natural experience, from

my sensible limitations.  Thus, if you take away
my natural body, leaving no similar body in its
place, you take away my sole ground of consciousness, the sole basis of my experience of the *me*, for
I become conscious, or say *me*, only by virtue of
my bodily constitution.

But now how would it do for God's creature to
be without selfhood, without any consciousness of
the *me*, that is without life?   For life *is* life only
in the ratio of the intensity of the individual consciousness, or of its own essential freedom.   Why
evidently it would do miserably : that is to say, it
would degrade the creature—I was going to say—
to a stone.   But the stone enjoys an *inert* consciousness, or discloses the *me* under a form of
inertia, and hence would still prove too flattering
a similitude of the creature thus viewed.   In truth
the creature, in the case supposed, would be a
strict zero infinitely below the mineral form even :
and his Creator by implication would descend to
the same level of nonentity, for there must always
be an exact ratio between Creator and creature.

It is, accordingly, a fundamental postulate of all
true or recognizable creation, that it be a living
one, that the creature be endowed with selfhood
or consciousness, and so placed in some rational
proximity to his creative source.   For it is the
precise peculiarity and perfection of the creative
name or quality, that He is infinite or unrelated,
that is to say, *without community of being*.   And

selfhood images this creative peculiarity, this Divine perfection, putting the creature so far as he realizes it into exact correspondence with the Creator, and making Him a full participant or subject of the Divine infinitude.

But how shall the creature realize this selfhood in any degree, seeing that he is absolutely destitute of it? It is evident from the bare statement of his creatureship, that he is *intrinsically* void of life or selfhood, that his very nature is *not to be*. How then shall he surmount this intrinsic or natural destitution, and so arrive at selfhood, or conscious existence? This is the question. On the one hand, it is clear that he cannot be a creature of God, save in so far as he possesses selfhood : on the other, it is clear that he is intrinsically, or by nature destitute of selfhood. And the problem is to reconcile these two propositions, or to shew this intrinsic natural destitution of life giving place to the amplest and eternal exuberance of it.

But now does it not irresistibly follow from these premises, that creation is nothing more and nothing less than a process of redemption? Does it not follow, in other words, that all true or Divine creation rigidly consists in giving the creature redemption from his own nature, that is, in endowing him with selfhood? Bethink yourself. We have just seen that God's true creature is bound by the necessity of his creature-

ship to possess selfhood or conscious life. And
that same necessity binds him to be intrinsically
void of life or selfhood, binds him to be in himself, or absolutely destitute of consciousness,—ob-
liges him, in short, as to his own nature *not to be*.
Do you not see then at a glance that God creates
man, or gives him selfhood, only by giving him
deliverance from this intrinsic void, only by redeeming him from this natural destitution? Of
course you do, and I will therefore take the
question for granted. But if creation consist in
delivering the creature from his own nature, or
endowing him with selfhood, then it becomes also
instantly clear that unless we turn creation into a
mere abstraction, or verbal juggle, the creature is
bound to have a natural or finite projection, as
well as a spiritual or perfect one,—is bound to
experience a *quasi* or phenomenal existence, as
well as a real or redeemed one,—is bound in short
to know himself in an *uncreated* state, so to speak,
as well as in a created one. I say this duplex
consciousness is obligatory, because otherwise
creation would not be a fact, but only a verbal
fiction. God creates me or gives me being only
in so far as He redeems me from my *natural* destitution, that destitution which stands in mere
*community* of existence : and He redeems or lifts
me out of this community by giving me selfhood,
which is individual freedom or expansion. But
obviously I shall have no power to appreciate or

even accept this Divine boon, unless I am able to contrast it with my native destitution: unless, in other words, I first experience a certain *community* with my kind, which, finiting me on every hand, and under the sensible or outward show of life filling my bosom with the spiritual and inward consciousness of death, bids me aspire with invincible yearnings after the real and imperishable life that comes from God alone.

Thus Nature avouches itself an inexpugnable necessity of the Divine creation. The natural world is implied in the spiritual world, just as the foundation of a house is implied in the superstructure, or the shell of a nut implied in the kernel. That is to say, it exists not for its own sake primarily, but for the sake of the use it promotes to a superior life. The foundation of a house may *within its own limits* be very commodiously disposed: it may be extremely well heated, lighted, and watered; may display unequalled kitchen, laundry, and dairy resources, and contain comfortable accommodation of all sorts for the servants: but clearly it will be what it is only with reference to the superstructure. Because the house itself, as to its magisterial portion, is of an uncommon excellence, it demands a corresponding excellence in these its subordinate or ministerial parts. So also the shell of a nut may be thick or thin, hard or soft, rough or smooth, but whatever be its concho-

logical peculiarities, they will be altogether determined by the necessities of the interior fruit or
kernel.   Precisely so with the natural life.   It
may exhibit *within itself* any amount of splendour
and comfort, but it is none the less tributary on
that account to the superior world.   In truth, its
proper splendour and wealth will accrue exactly
in the measure of its subserviency to the demands of the higher life, that is, in the ratio of
its use.

It is on this point precisely that the orthodox
theology and philosophy signalize their inherent
incapacity to furnish us with a true doctrine of
Nature, or confess themselves utterly unscientific.
Our popular theologians and philosophers have
no idea that nature is but a correspondence by
inversion of spirit, just as the foundation of a
house is a correspondence by inversion of the
superstructure, as the shell of a nut is an inverse correspondence of the kernel, or the outside
of a glove an inverse correspondence of its inside.   On the contrary, they deem the natural
world to possess an independent existence, to
exist for its own sake, or constitute its own end;
and consequently they have declined, both classes
alike, into mere Naturalism.   The current theology and philosophy are both alike naturalistic,
whence we now have Unitarianism as the only
vital theologic doctrine extant, and Atheism or
Pantheism as the only vital philosophic doctrine.

We live under the Iscariot apostolate. The star of the forlorn Judas culminates at length in our ecclesiastical horizon, and we have little left to do but to burst asunder in the midst, or resolve our once soaring Divine hopes into the mere poetry and sentimentality of nature. There is scarcely a theologian in the land who does not tacitly regard the soul as a thing; and he who was recently the idolized chief of your philosophic hordes, habitually regarded infinitude as identical with the totality of space and time, and under that conception very properly execrated it as an " IMBECILITY of the human mind." One can easily imagine the very inconsiderable figure which Sir William Hamilton, supposing him to be the same man intellectually that he was a few months ago, makes among the unsophisticated—or rather the desophisticated—angels. Looking as he did upon the infinite as meaning in respect to space the all of space, and in respect to time the all of time, he must of course either deny God altogether (that is, acknowledge Atheism to be the final flower of Philosophy), or else identify Him with the natural universe (that is, acknowledge Pantheism to be the upshot of Philosophy). And both the Atheist and Pantheist, according to Swedenborg's lively daguerreotypes of trans-sepulchral existence, experience an immense deal of pulmonary oppression in angelic atmospheres. In truth I am afraid that those innocent angels will

turn out rather "slow" people to most of our
grandees, sacred and profane: they are so little
skilled in our operose intellectual gymnastics, and
acknowledge God so much more as little children
than as turgid and palpitating athletæ. However there may be new-fashioned angels as well
as new-fashioned men, nay there *must* be: and
we will not doubt therefore for an instant that
every honest inquirer will yet find himself, in
spite of any amount of latent atheism and pantheism, in the divinest possible circumstances.
Cheerfully according the theologian and philosopher then so excellent a look-out, let us leave
them for the present to ask, in the light of all
that has gone before, what is the precise intellectual infirmity designated by Naturalism? When
we say that Naturalism is the disease of our current divinity and metaphysics, what do we mean
to indicate by that word?

We mean to indicate the prevalent habit of
regarding Nature as an absolute or positive existence, and not as the mere inverse and negative
aspect of spirit. Naturalism limits the reason by
the senses; it accepts as final the testimony of
sense affirming the identity of being and appearance, substance and shadow, or what is the same
thing, affirming that all real existence is constituted of space and time. The consistent naturalist says in all his thought, " I *am* inwardly and
spiritually what I outwardly and physically *ap-*

*pear :* that is to say, as I am naturally distinct from and disunited with all existence, so am I also spiritually distinct and disunited : and distinct and disunited existences deny unity of origin, deny that one and the same Creator could produce so many divergent creatures."

Thus the naturalist limits the reason by the senses, or allows the phenomenal to dominate the real. Not seeing that Nature is but the inverse of spirit, that natural variety and difference are but the inverse correspondential expression of spiritual unity, he allows the former to dominate the latter, and conceives of the soul as existing only in bodily conditions, the conditions of time and space. Going to the mirror for instruction as well as information, for wisdom as well as knowledge : asking it, not how things *appear* to what is below themselves, but how they *exist* to themselves : he becomes hopelessly duped, and ends by not knowing his right hand (spiritually) from his left, or his heels from his head. Misled by what Swedenborg calls the sensuous *lumen,* the mere light of Nature, we invariably immerse spiritual existence in material dimensions, or subject it to time and space. Denying the unity of the soul in God, thus the unity of humanity, we split the spiritual creature up into as many conflicting and independent and selfish souls, as Nature exhibits of bodies. In short, the naturalist, instead of making nature and spirit twin aspects of one and the same consciousness, as that consciousness is viewed either subjectively or objectively : instead of seeing both the natural and the spiritual universe alike included in the unity of the conscious *me*, both alike pervaded by the unitary human soul, both alike embraced in humanity, in short : gives them the reciprocal independence and obtuseness of two peas, and so leaves them utterly and eternally destitute of rational accord.

Yours truly,

————.

# LETTER XII.

*Paris, Dec.* 15*th*, 1856.

MY DEAR W.,

' THE incomparable depth and splendour of Swedenborg's genius are shewn in this, that he alone of men has ever dared to bring creation within the bounds of consciousness—within the grasp of the soul. He alone has dared to give to Nature human unity,. to endow it with the proportions of man. This is the fundamental distinction between his genius and that of all our other great writers, that while they, by *exteriorating* object to subject, Creator to creature, God to man, materialize man's motives, and so construct a grossly sensual Theology, and an utterly selfish Ethics; he in *interiorating* object to subject, God to man, spiritualizes man's motives, and consequently constructs a Theology which places God exclusively within the soul, and an Ethics whose sanctions lie in the demands of our endless spiritual development, and no longer in the arbitrary pleasure of any foreign power. As it is impossible to comprehend the laws of creation without a clear per-

E 3

ception of the rational truth on this subject, I
shall make no apology for dwelling upon it a
while longer.   I simply seek to familiarize you
with the fundamental truth of Swedenborg's system, which is that God is essential man, and that
all creation consequently is in human form, being
everywhere pervaded by consciousness more or
less perfectly pronounced.

In a former letter I shewed you that the *me*
absorbs the whole realm of the finite, the domain
of sensible experience, the *outer* sphere, so to
speak, of consciousness.   The *not-me* equally absorbs the realm of the infinite, the domain of
spiritual experience, or the *inner* sphere of consciousness.   Consciousness forms the dividing and
yet uniting line between infinite and finite.   It is
the hyphen which separates yet unites the object
and subject, the not-me and the me.   Whatsoever
is on the *hither* side of consciousness, whatsoever
is sensibly discerned as mineral, vegetable and
animal, is finite and falls below the *me*.   The *me*
dominates it.   Whatsoever is on the *thither* side
of consciousness, whatsoever is spiritually discerned, as goodness, truth and beauty, in short
character, is infinite and falls above the *me*.   The
me aspires to this infinitude, cultivates it, wor-
ships it.   Consciousness, or life, unites this higher
and lower realm, giving us the *beautiful* mineral,
the *graceful* shrub, the *gentle* animal, the *good*
man.   The grammatical adjustment of adjective

and substantive is only a formula of the copulation which all life or consciousness implies between object and subject, between infinite and finite, between the not-me and the me. This is the invariable meaning of consciousness: *the copulation of an interior object with an exterior subject; the marriage of a universal substance with a specific form.* Wherever there is organized life or consciousness, there is the coupling or congress of an interior infinite object with an exterior finite subject: the marriage of a universal and invisible substance with a specific and visible form: the life or consciousness being high or low, rich or poor, human or inhuman, precisely as this marriage is more or less perfectly pronounced. In short, consciousness or life invariably asserts the union of a universal interior substance with a particular exterior form, the life being more or less perfect, that is *human,* just as the union in question is more or less complete in the subject, that is to say, just as the individual subject is capable of *universalizing* himself, of adjusting himself to universal relations.

According to this definition, man is the highest form of consciousness, because in him alone is the individual element proportionate to the universal. Man is the only universal form. He stands related to universal nature, on the one hand, by what he possesses in common with it, and to God on the other, by what he possesses over and above such natural community. He is related to the mineral forms of nature, by gravitation and consequent inertia: to its vegetable forms by sensation and consequent growth: to its animal forms by volition and consequent motion; while he alone claims relationship with God, or the infinite, by what he alone possesses; namely, spontaneity, or the power of unforced individual action. The animal does not originate his own action, that is, is destitute of spontaneity. He acts wholly from the control of his nature. And man, so far as he is animal, does the same thing. But in so far as he is man, he acts from taste or individual attraction, that is to say, originates his own action, or exhibits spontaneity. The mineral is the least perfect or human form of life, because it exhibits the universal element in such superior force or volume to the individual one. Gravitation, inertia, rest in space, is what all existence possesses in common. And yet the mineral which expresses this common characteristic, is the lowest form of existence. Indeed, philosophers utterly deny life or consciousness to the mineral form, because the individual, or formal and feminine, element, is so inferior in it to the universal, or substantial and masculine, element. The latter element almost swallows up the former, nearly reducing the mineral to what the philosophers call " a simple substance."

But this denial is premature. Consciousness

belongs to the mineral realm as truly, though not
so distinctly, of course, as to the vegetable and
animal.   It is the most diffused or common form
of consciousness, and therefore the least obvious
to human apprehension.  For man being the most
distinct or pronounced form of life, that is to
say, harbouring the universe in his private individuality, is at the utmost possible remove from
communism, and hence finds it very difficult to
appreciate or even acknowledge a life which simply expresses that.   But nevertheless the fact is
so.   Mineral life or consciousness is no doubt
very base compared with the human, but still it
is real.   It is, no doubt, very base and imperfect
compared even with the vegetable or animal.   For
vegetable growth, and animal motion, are much
less diffused or communistic forms of life than
mineral inertia or rest.   But imperfect as it is,
it is still life.   It is indeed the base of all higher
life or consciousness, vegetable growth and animal
motion being only modifications of that base
operated by the advancing life of nature, by the
exigencies of the perfect or human form.   The
perfect or human form is that which exactly
unites or marries what is universal and what is
individual, the sympathies of every well-developed
man relating him to the entire universe of being.
Mineral life is the first step towards this perfected life.   It is the arrest of chaos.   Originally
or in the uncreated state of man, so to speak,

Creator and creature, God and man, are undistinguishable one from the other. And Nature in her earliest or fluid beginnings does but exactly symbolize this indistinction. All things are then chaotically blent. Confusion reigns: that is to say, there exists a sensible fusing together and indistinction of all forms of individuality, of all forms of life or consciousness, under strictly universal forms. And the entire scope of what we call *history* is to reduce this chaos to order, to lift up this sobbing and prostrate universe into beautiful and joyous and individual form, to train this mute and melancholy and boundless nature into the free and glorified lineaments of human personality or character.

The mineral form then is the earliest or lowest evolution of the me. It is the me in an intensely inert state, in a passive state or state of rest simply. It is the me getting place or position first, in order to its subsequent experience of *growth* in the vegetable form, *motion* in the animal, and *action* in the human form. We may say that it is the me in a fœtal state, conceived but not yet born. Mineral life bears the same relation to vegetation and animation, that the fœtus bears to fully developed bodily life. Being destitute as yet of sensation, of sensibility to outward existence, it is of course devoid of visible form or individuality, and loses itself in the womb of the common mother. But the phenomena of crystallization shew that the life-process is going on all the while not less really though invisibly, or within the still enveloping womb of Nature, so that at least very marked *tendencies* to specific form result, as we see in the characteristic differences of iron and sulphur, alum and arsenic, gold and lead, silver and copper. Were there no observable differences in these things, did they not exhibit each a different relation towards the common mineral life, which is inertia or tendency to rest, we could never have named them. But if you admit mineral differences together with a universal mineral nature, you admit mineral life or consciousness. For life or consciousness means nothing else than the union of a common nature with a specific form.

But I will return to this topic in another letter.

Yours truly,

————.

# LETTER XIII.

*Paris, Dec. 20th,* 1856.

MY DEAR W.,

TAKING humanity in its largest scope, we may say that the mineral kingdom forms its osseous structure, or gives it stability: the vegetable kingdom forms its fleshly structure, or gives it sensation and consequent growth: the animal kingdom forms its nervous structure, or gives it volition and consequent motion: while man is the regal selfhood to which all these kingdoms tend, in which they all culminate, and by the mediation of which, all finite as they are, they connect with the infinite. Thus we may place mineral existence very low in the scale of humanity, but we have no right to exclude it from that scale. The characteristic of mineral life is inertia or rest, while that of perfected human life is freedom or progress. But how is progress possible without a starting point, and a starting point moreover perpetually renewed? Or how is freedom conceivable without the contrast of some fixity? What *is* freedom indeed but an eternal escape or rising away from all fixity and routine? Thus, if you deny the mineral element in humanity, you deny its principle of identity, or that thing which houses it under all changes of sky, keeping it fresh and sweet and immutable through eternal years, as when it knew only the maternal embrace. It is, in fact, as I have already said, the fœtal condition of humanity. It is the precise form of life exhibited by the human fœtus, while it absorbs from the enveloping heavens and earth of the maternal bosom, those increments of corporeity which shall one day extrude it from that tender abode, and impel it through all the vicissitudes of vegetable growth, animal motion, and human action, into full acknowledgment of the infinite, into final union with God. It is not so dignified a form of life then, let us grant, as many others, but our own consciousness fully attests its reality. We have no need to descend into the bowels of the earth, to ask the gold, the sulphur, or other mineral existence, whether it have any and what form of life or consciousness, for we ourselves involve the mineral consciousness, and can accurately describe it. We often experience the mineral life or consciousness, as, for example, when we fall from a height to the ground, or merely from a perpendicular to a horizontal position. Evidently that thing which my body possesses *in common* with all bodies, namely, gravitation, and which makes it incessantly *tend*

with all bodies to a common centre, is what alone
produces the fall. If you take away my mineral
characteristics, or that most diffused and common
nature which links me to all other bodies, you
take away my liability to fall. You destroy me
as a subject of the mineral nature. But now in
these mineral experiences of ours, the *me* does
not cease to exist. It simply undergoes a transient degradation, and from being intensely exerted, becomes suddenly intensely inert. What
is human in us, and what should ensure us the
obedience of all lower nature, becomes as it were
inverted and threatened with immersion or suf-
focation under lowest forms. For in this mineral experience, we keenly perceive that the
*me* or individual and feminine element, instead
of *concurring* with the universal and masculine
element—instead of bringing forth fruit to it
instinctively like the vegetable, or voluntarily like
the animal, or spontaneously like man—is dominated or coerced by it, and hence yields it in
place of the hearty co-operation which the wife
yields the husband, only the grudging, and struggling, and rebellious submission the slave gives
the master. It is an *in*-ertion of the me, so to
speak, or the me vigorously resisting the dominion of mere natural community, instead of an
*ex*-ertion of it, which supposes nature already
separated and subordinate. It is the me getting
body, and this rudimentary body, as we know

from its analogy in the fœtus, is long destitute of
perfect human form. It is the very core and
back-bone of the me that is forming, the very
centre and focus of consciousness, and the less
vital but more demonstrative periphery, like the
still undeveloped extremities of the fœtus, is tenderly latent in that. This is all the difference.
The mineral life is just as real as the vegetable or
the animal life, only it is as the life of the most
vital *viscera* of the body which shun the eye,
compared with that of the surface and extremities.
The vegetable and animal forms of life are only
so many—not more real—but more sensuous
evolutions of the me which is latent in the mineral. What is inertia in the mineral, or the
simple *statics* of the me, becomes sensation in the
vegetable and volition in the animal, which are
*dynamic* declarations of the me. Did the me not
first wear this form of inertia,—this form of re-
sistance to the overwhelming fluidity and community of nature, it would never burst forth in
the higher or vegetable form of sensation. The
mineral inertia marks the initiatory protest of the
me against total community of nature: it is the
beginning of that absorption which all mere community is bound to undergo into beautiful and
distinctive form: vegetable growth, animal motion, and human action, only record the successive triumphs in which that initiatory protest
ends. How beautiful the phenomenon of vege-

table growth! Here we see this chaotic, this
communistic and formless nature, sopped up, so
to speak, and trained into forms of exquisitely
modulated variety: in other words, we see this
vast and vague and overpowering universality becoming personal and human, becoming resolved
into clearer and clearer individuality. The animal life is only a more advanced evidence of the
same process, is only a still more vivid picture of
the inevitable marriage between infinite and finite,
while in man the marriage culminates, and we
see all nature at last joyfully acknowledging her
sovereign Lord, or Divinity perfectly glorified in
Humanity.

I hold, then, that there is no such thing as
unorganized, unconscious, or dead existence; but
I hold this incidentally to my main proposition,
which is that all life is a form of consciousness,
of a joint or composite self-knowledge, implying
the union of an interior object with an exterior
subject, the marriage of a vivifying succulent
nature with a dependent specific form. It is this
marriage-union which invariably determines what
we call the selfhood, or personality of the subject.
Hence the mistake of the Idealist in denying
nature a soul of her own, and making her sensible qualities inhere in a foreign subject. The
Idealist denies the substantiality or absoluteness
of Nature within her own plane, and hence
affronts the universal conviction of mankind

which attributes to nature a personality quite independent of her sensible properties. This personality is not a phenomenon of sense. You cannot resolve it into any sense, nor any number of senses. The selfhood of the pear-tree does not lie in the form of the tree as determined by my eye, nor its odor as determined by the nose, nor its solidity as determined by the touch, nor yet by all these things put together: but simply in its power of prolification, that is to say, its power to propagate its own nature. So also the selfhood or individuality of the horse does not consist of his sensible properties, or those things which relate him to our intelligence, but of his natural faculty of reproduction, or what is the same thing, his subjection to the motions of his invisible nature. In short, all life, all personality, all character, dates from the union in question, confesses itself the offspring of a marriage between an objective controlling nature, and a subjective obedient form. Undoubtedly, as I have already observed, the resultant form is low or high, poor or rich, exactly as the marriage is ill or well pronounced, that is, as the formal and feminine element is *freely* instead of *servilely* related to the substantial and masculine element. But whatever variations characterize natural forms, they all *equally* confess themselves the progeny of a marriage between an interior controlling nature and an exterior submissive form.

But now you know that a very marked difference obtains between the human form, or personality, and that of mere mineral, vegetable, or
animal existence.   It is true, that the human
form, equally with these lower forms, confesses
itself the offspring of a marriage between a common nature and a specific subject.   Rather let
me say, that the human form makes this confession with supreme distinctness : for the marriage in question is so much more emphatically
pronounced in that form, that it may be said to
be comparatively unpronounced in every other.
For example, every mineral, every vegetable, and
every animal form, while they exhibit great reciprocal diversity, yet sink into the same undistinguishable level before the universality of man.
Take the highest of these forms, the animal.
Compared with the vegetable and mineral forms
of life, the animal ranks heaven high by the bare
fact of will and consequent motion.   But when
you view the animal form in itself, when you ask
how it stands related to the universal life, you instantly see that it has no individuality answering to that universality.   In short, you find no
unitary animal form below the human.   The lion
is out of all unison with the cow, the fox with the
sheep, the serpent with the dove : look where you
will, diversity not unity, discord not concord, is
the law of animal life.   One animal preys upon
another ; one half of the animal kingdom lives

by destroying the other half. Now man, so far
as his natural form is concerned, resumes all these
distinctive differences of the lower natures, and
fuses them in the bosom of his own unity. He
is not only devouring as the fire, and unstable as
the water: he is fixed as the rock, hard as the
iron, sensitive as the flower, graceful and flowing
as the vine, majestic as the oak, lowly as the
shrub. But especially does he reproduce in himself all the animal characteristics. He is indolent
as the sloth, he is busy as the bee, he is stupid as
the ox, he is provident as the beaver, he is blind
as the bat, he is far-sighted as the eagle, he
grovels like the mole, he soars like the lark, he
is bold as the lion, timid as the fawn, cunning as
the fox, artless as the sheep, venomous as the
serpent, harmless as the dove: in short, all the
irreconcileable antagonisms of animate nature
meet and kiss one another in the unity of the
human form. It perfectly melts and fuses the
most obdurate contrarieties in the lap of its own
universality. It is this universality of the human
form which endows it with the supremacy of
nature, and fits it to embosom the Divine infinitude. Because it adequately resumes in its own
unity the universe of life; because it sops up, so
to speak, and reproduces in its own individuality
all mineral, all vegetable, and all animal forms, it
claims the rightful lordship of nature, or coerces
nature under its own subjection. Thus the marriage I speak of is perfectly ratified only in the

human form, because in that form alone does the
feminine or individual element bear any just ratio
to the masculine and universal one. In short,
man is the sole measure of the universe, because
he alone combines in the form of his natural
individuality every conceivable characteristic of
universal life.

It is clear, therefore, that although man may
be said to be subject to his nature just as truly as
the horse and the rose are subject to theirs, yet
the human nature is of such a measureless scope
and dignity, claims such a universal pith and
variety, as to lift all its subjects at one and the
same *coup* out of the realm of physics, and bring
that realm within the invincible grasp of their
subjectivity. Physics ends precisely where man
begins. Mineral, vegetable, and animal exist only
to endow his commanding individuality, only to
universalize his form, only to give him a basis
broad enough to image the Creator's infinitude.
He is the dazzling blossom of the universe, the
peerless fruit by whose interior chemistry unripe
Nature ripens all her juices to gladden the heart
of creative Love. Thus by the very law of its
creation all nature aspires to the human form,
confessing itself the mere blind type and stutter-
ing prophecy of that unmatched perfection. In a
word, nature acknowledges herself contained in
man, cheerfully dons his livery, and obediently
reflects his life.

We can have no difficulty now in estimating

the exact difference between man and all lower forms of life. He, at his lowest, is a universal form of life : they, at their highest, are only partial forms. In this distinction is expressed all the distance between man and nature, between human history and mere animal or vegetable growth and decay, between man's eternal progress and nature's eternal immobility, between the starry splendours in short of human society or fellowship, and the dull ungenial fires of mere brute community. For it is this difference which makes man a fit subject of God, and suspends nature's alliance with Him only on man's mediation. But this subject can hardly be broached short of another letter, and for the present I subscribe myself,

Yours truly,

———.

# LETTER XIV.

*Paris, Dec. 25th,* 1856.

MY DEAR W.

I HAVE now virtually answered the question which in my ninth Letter I represented you as asking, namely : how the rectification of our intellectual methods recommended by Swedenborg, avails to give us a right apprehension of creation, and promote a truly scientific cosmology. That is to say, I have alleged nearly all the considerations which determine the answer to that question, and little remains but to draw the answer out in legible characters. But before doing this, I want to fix your attention upon what I have already discussed, but what I have not perhaps sufficiently insisted upon, and that is, the immense peculiarity or distinction of the human form. I have shewn you that creation is bound by the creative unity to wear a unitary form, and that this unitary form is that of humanity. Now what is the human form as distinguished from all lower forms? What is the *distinctive* form of man, his form as *contrasted* with the animal or vegetable or mineral form?

In one word, it is a *spiritual* form, and if you now ask me what I mean by a spiritual form, I reply that I mean a form which is enlivened or empowered exclusively from *within;* a form, in other words, whose activity invariably reflects its own free spirit or selfhood, and disavows all outward constraint. This is the only form adequate to the Divine subjection, adapted to the Divine inhabitation. God does not create life of course, (for life is uncreated), but only *forms* of life, to which He incessantly communicates life by His own spiritual indwelling. Inasmuch, therefore, as God is life itself, inasmuch as His infinitude or perfection flows from His character, derives from Himself, and is accordingly altogether spiritual, disclaiming all outward genesis, you are bound of course to exact a responsive image in His creature. You are bound to exact a creature, the form of whose life shall be intensely spiritual, and natural only in furtherance of that. God, being the infinite or perfect spirit He is, cannot create— cannot give being—to a lower or unspiritual style of life. A lower style of life in the creature would argue a finite Creator, would exclude a perfect Divine original.

Now spiritual life, that style of life which expresses the selfhood or freedom of the subject, and which consequently characterizes him, is emphatically *human* life. This is precisely the definition of humanity, namely, the power of free or characteristic action, the faculty of spontaneous life, the ability to obey in all cases one's own taste or attraction in opposition to physical or moral constraint. We feel that life is human in proportion as it is free, that is, in proportion as its merely external, natural, or communistic element freely subserves its internal, spiritual, and private element. This is the reason why we instinctively dislike Pharisaism, or an inflated state of the moral consciousness. The Pharisee is filled with a sense of his own merit, because he has *denied himself*, because in obedience to some law of the community, he has refrained from doing as he otherwise would have done: and we instinctively feel that a man should be ashamed of a virtue which proceeds only upon his own suppression, which exists only in so far as he renounces his own freedom. We have, indeed, more hope of the publican and harlot than of such a man, because so long as this man prides himself upon what really avouches his own degradation, he puts himself out of all true relation to conscience, and instead of finding it a minister of death, a flaming sword keeping the way of the tree of Life, discovers it to be a Divinely blunt and bland witness of his own righteousness, and hence feels himself very properly acquitted of any obligations to a coming Divine righteousness: while the publican and harlot, on the contrary, being denied all moral righteousness, all righteousness in themselves as contrasted with other men, are extremely likely to lend a gratified ear to every Divine promise in that direction. In a word, the desperate evil of Pharisaism is that it leaves one content with a finite life or righteousness, such as flows from a completely servile relation to the obligations of natural community, and so disqualifies him to appreciate that infinite life or righteousness which stands in a relation of free superiority to our nature, and of consequent intimate conjunction with God. The publican and harlot are comparatively void of this disqualification, and hence lend themselves far more cordially to Divine medication. *Verily, verily, I say unto you, the first shall be last, and the last first.*

However this may be, I repeat that it is the precise distinction of human life or action that it is free, that it is spontaneous, that it always dates from *within* the subject, that it obeys only an interior motive, and disclaims both physical and social coercion. Eating and drinking to supply the needs of one's nature, are not distinctively human attributes : they belong to man on his animal side, or as he stands related to his physical organization. He is not free accordingly to forego his physical action, nor consequently is he free in producing it. He would starve if he did not produce it. His moral activity is just as servile and uncharacteristic. It belongs to man on his social side, or as he stands related to his fellow-man, and he is not free to forego it. His relations to his family, to his tribe, to the community in which he lives, impose certain obligations upon him by the promise of certain rewards, or at all events the menace of certain punishments, and he must either obey or suffer. He must do whatever the law of his community prescribes to him, or he must be socially banned and punished, perhaps killed. Thus neither the natural nor social side of man, neither his physical nor his moral action, reflect his distinctively *human* character. The natural and social side of man, the realm of his relations to nature and society, is the realm of LAW or fixity, in which he is only apparently free, while *really* he is in invincible bondage. The spiritual or interior side of man, the realm of his relations to God, what we may properly call his supernatural realm, is the sphere of his veritable LIFE, in which he is not only apparently, but also most *really* free. To employ Swedenborg's terminology, delight is the essence or motive of this life. Whatsoever the subject does he does from the pure delight of doing it, from attraction, from inward taste, or spontaneity. It is as Swedenborg phrases it, a life of cordial or spontaneous *use*, meaning thereby an æsthetic life, a life of free productivity, utterly ignoring the stimulus either of physical necessity or moral obligation.

Surely no argument is needed to sustain these

positions. Clearly if the distinctive life of man lay in his relations to nature, then we should never see him renouncing natural obligations, the obligations he owes to his own body. We might, it is true, see him dying just as the animals die, but we should never see him committing suicide. And so also if his distinctive life were moral, or lay in his relations to his fellow-man, we should never see him capable of renouncing those relations, or violating the duties they impose. We might find him perishing under the exactions of society, but we should certainly never see him as now contemning his social obligations, and deriding the penalties which society thereupon affixes. Doubtless no one can act from a superior plane to that of his own life. If, then, nothing is more common than to see men renouncing their physical and social subjection, we must allow that they do so only in obedience to the instincts —blind it may be and unenlightened, but still most real—of a life which is yet more truly their own. Human life, in a word, is not primarily natural, does not acknowledge a physical origin : or we should have no suicide. Neither is it primarily moral, flowing from a social origin: or we should be destitute of moral evil. Nothing has been more common in the past than to see man obstructing, mortifying, harassing his natural body with a view to some ideal or spiritual end. It is as if the more vital life within were in such haste to

come to consciousness, that it could hardly forbear to usurp the natural organs.   And certainly
nothing is more common now-a-days than to see
men obstructing and embarrassing their moral
life, the life which flows from their relations to
society, with a view to some interior spiritual
ease.   The poor struggling wretch is all unconscious of the sacred instincts which at the bottom
animate his perverse activity: he can tell you
nothing, because intellectually he knows nothing,
of the profound human want, the want of freedom, which like the breath of the whirlwind
hurries him on : on the contrary, he will very
probably accept your otiose and overbearing
hypothesis of the case, and say that he does evil,
simply because he is an evil man unworthy of
human love, worthy rather of human scorn, worthy only of Divine and human vengeance.   But
all this is sheer insanity on his part and ours.
There are no fundamental differences in men.
All men have one and the same Creator, one and
the same essential being, and what formally differences one man from another, what distin-
guishes hell from heaven, is that they are differently related to the Divine *natural* humanity, or
to the life of God in nature, which is a life of
perfect freedom or spontaneity.   In that life self-
love freely subordinates itself to neighbourly love,
or promotes its own ends by promoting the wel-
fare of all mankind.   But so long as this life is

wholly unsuspected by men, so long as no man dreams of any other social destiny for the race than that which it has already realized, and which leaves one man out of all fellowship or equality with another, self-love is completely unprovided for, except in subtle and hypocritical forms, and is consequently driven to these disorderly assertions of itself by way of actually keeping itself alive. Thus, whether he is unconscious of the truth or not, no man is evil save for want of free development, save for want of a closer walk with God than the existing Church and State agree to tolerate. The liar, the thief, the adulterer, the murderer, no doubt utterly perverts the Divine life which is latent in every human form: he degrades and defiles self-love, in lifting it out of that free subordination which it will evince to brotherly love in the Divine *natural* man: but he nevertheless does all this in the way of a mute unconscious protest against an overwhelming social tyranny, which would otherwise crush out the distinctive life of man under the machinery of government and caste. Accordingly, I am profoundly convinced that if it had not been for these men, if we had not had some persons of that audacious make which would qualify them to throw off their existing social subjection, and so ventilate, even by infernal airs, the underlying life and freedom of humanity, that life and free-

dom would have been utterly stifled, and we
should now be a race of abject slaves, without
hope towards God, without love to our fellow-
man, contentedly kissing the feet of some infallible Pope of Rome, contentedly doing the bidding
of some unquestionable Emperor of all the Rus-
sias.   These men have been, unknown to themselves, the forlorn hope of humanity, plunging
headlong into the unfathomable night, only that
we by the bridge of their desecrated forms might
eventually pass over its hideous abysses into the
realms of endless day.   Let us, then, at least
manfully acknowledge our indebtedness to them :
let us view them as the unconscious martyrs of
humanity, dying for a cause so Divinely high as
to accept no conscious or voluntary adhesion, and
yet so Divinely sure and sweet and human as ultimately to vindicate even their dishonoured memory,
and rehabilitate them in the love and tenderness
of eternal ages.   In short, let us agree with Swedenborg, that odious and fearful as these men
have seemed in merely celestial light, they have
yet borne the unrecognized livery of the Divine
NATURAL humanity, and will not fail in the end
to swell the triumphs of His majestic patience.
And this simply because by an undying Divine
instinct, under every depth of degradation celes-
tially viewed, they have always been true to themselves, feeling themselves to be men and not
devils, and over their scarred and riven legions

have ever indestructibly waved the banner of a
conscious freedom and rationality.*

But dismissing argument, I assume as indisputable my original proposition, which is, that
the human form is essentially free or spontaneous.
It is free or spontaneous because, as we have seen,
its universal force freely yields to its individual
force, or all that is natural and common in it
serves only to promote what is spiritual and pri-

---

\* Let me moreover remind the reader, here, that the hells, as we
have seen in Letter IV., are purely excrementitious. Now in Swedenborg's time, the scientific appreciation of manures, in redeeming
worn-out or vastated soils, was almost comparatively unknown ;
and he had accordingly scarcely any analogy to guide him as to the
altogether splendid and benignant uses the hells promote in the
sphere of the redeemed natural mind, or in obedience to the
DIVINE NATURAL MAN. I have myself known thick-headed
Dutch farmers in the valley of the Mohawk, in the State of New
York, to sell off their homesteads, because there was such an accumulation of manure in their barn-yards as to render their barns
and granaries almost inaccessible. Feeling the rich inheritance to
be only an incumbrance and curse, because they knew nothing of
the endless Divine blessing with which its steaming and odious
bosom was fraught, they concluded to sell their exhausted farms
rather than remove their barns, and with the proceeds go to buy new
ones in the virgin soils of the West. Yet all around these precious
heaps the grass grew so luxuriantly, that you would suppose
stupidity itself would take a hint. But our spiritual husbandmen
display the same obdurate contempt for the infernal element in
humanity, nor ever dream that such priceless Divine renovation for
the exhausted natural mind of man is stored away in those now
nauseous and festering because useless forms, to which we give
the generic title of the Devil. But the subject is too vast for a
note.

vate. Hence alone is it a fit form of the Divine
infinitude: hence alone does man avouch himself
a proper subject of God. For the human form
thus asserted exactly images God. The necessary
conception we have of God is, that He is a universal Creator, or in plain English, that He alone gives
being to all things. Thus we have individuality
(God Himself), and universality (all things), both
involved in the idea of God as a creator; and
then, further, we have the universality confessing
itself as subordinate to the individuality. All
things derive from God. Obviously then the true
creature of God must exactly image this creative
perfection. He can neither fail to exhibit in his
own consciousness the two elements of individuality and universality, nor yet to exhibit the latter
element in a relation of perfect subjection to the
former. He cannot avouch himself a creature
otherwise. We cannot regard him as really
created until we recognize him in this form,
because, manifestly, any other form would be
inadequate to God's inhabitation or indwelling.
Creation is not the production of new being, but
only the formal manifestation of a being which,
being infinite and eternal, is never new and never
old, which is without beginning and without end,
and which, therefore, utterly ignores those laws
of time and space on which its mere manifestation is contingent. In other words, God creates
or gives us being only in so far as He first gives

us form, only in so far as He first makes us images of Himself, which forms or images He may then fill with all the fulness of His life and delight as a river fills its banks, or the air fills the lungs. It is, therefore, simply absurd to talk of our being created, until we have realized this needful form or imagery.

It is quite common in loose discourse, no doubt, to confound creation, or the giving being to things, which is a purely spiritual process, with their making, or the giving them form, which is a purely natural process. It never seems to be popularly suspected that the two processes are so wholly discriminate as to be, in fact, reciprocally inversive. But all true Revelation rigidly proceeds upon such discrimination, as I shall shew you in my next.

Yours truly,

————.

# LETTER XV.

*Paris, Dec. 28th,* 1856.

MY DEAR W.

*In the beginning, God created the heavens and the earth, and the earth* WAS WITHOUT FORM AND VOID. Thus creating and making, or giving being and giving form, are two distinct things, and the mystic narrative of the six-days' work accordingly goes on to shew us the one becoming gradually proportionate to the other, until at last we reach the seventh, or perfected day, when we read: "And God blessed the seventh day and sanctified it, because in it He rested from all His work *which He had* CREATED TO MAKE." It is the unmistakeable mark of all true Revelation that it is constructed upon this rigid discrimination of being and form, of creating and making. All inspired writing involves a double sense: one spiritual, adapted to the rational eye exclusively; the other natural, adapted to the sensuous understanding. The internal sense alone is true, being full of the divinest and most ravishing particulars, just as a great muscle of the human body is full of minute fibres and fibrillas, too exquisitely individual to be discerned by the unassisted eye; while the external sense is made up of seeming or *quasi* truth, of truth in its most aggregated or common form, and bears, therefore, no more proportion to the vital Divine reality, than the gross muscle aforesaid does to the marvellous individual fibres of which it is the aggregation or accumulation. The spiritual sense gives us the eternal verity of creation, creation as regarded *à parte Dei:* the carnal sense gives us the form or appearance which this verity assumes to the consciousness of the creature. The spiritual sense alone claims objective truth: the latter possesses an exclusively subjective force. The one reveals the unchangeable being we have in God: the other tells us of the living evolution or formation which that being undergoes to our own consciousness.

Thus there is nothing arbitrary or irrational in this construction of the inspired cosmology. It takes place only because that cosmology really involves a far profounder philosophy of human life than has ever entered our best philosophic noddles to conceive. For the simple reason that we are created and not uncreate, our being must of necessity be entirely distinguishable from our form, must refer itself wholly to our Creator, while our form, on the other hand, is what identifies ourselves. " Creating" means *giving being,* causing *to be.* To create a thing is to

give it inward being or substance. "Making,"
on the other hand, means *giving form* or causing
*to appear*. To make a thing is to give it form
or phenomenality. Creating is a spiritual or internal process, having reference to the real, substantial, and objective nature of things. Making
is a natural or external process, having reference
to the phenomenal, formal, and subjective nature
of things. The former process would be known
to the creative personality alone, and the latter
fall exclusively within the created consciousness,
if it were not that the Divine mercy is too grand
to invest its creature with mere natural form
alone, and therefore aspires to communicate its
own deathless being to him as well. Thus
creating affirms the being or substance of things,
while making affirms their seeming, their appearance, their form. He who creates or gives being
to a thing, is Himself the substance of the thing,
and hence in so far as a thing should be created
merely, it would be undistinguishable from its
Creator. On the other hand, so far as a thing
is made or formed, it is individualized and discernible from its creative source. All making or
formation is development, is generation. A thing
is made or formed by being brought out of something else. Every thing that has form supposes
some precedent thing, or state, out of which it
grows, or by means of which it is made to appear.
The paper and the table upon which I write, the

pen and the hand that holds the pen, and the
body which sustains the hand, all these various
forms or phenomena of existence suppose some
precedent source, by virtue of which they themselves become manifest. They each suppose some
sensible background giving it objective unity with,
and subjective diversity from, all other existence.
This duality of existence enters into the very
conception of making or giving form, while the
conception of creating or giving being, on the
other hand, rigidly excludes duality, and shuts us
up to the Creator alone. Hence creation is never
a finite process, never falls within the laws of
space and time, that is to say, within the intelligence of the thing created. It is a truly infinite
process, pertaining wholly to the intelligence of
the Creator. The being of the statue is in your
mind; you create it or give it being in your
thought, whence, indeed, it may never come into
existence or form. But you make it or give it
form only in so far as you bring it out of the
rude marble, or separate it from the maternal
womb. Your formative power may never, indeed,
at any given time, equal your creative power : in
other words, the being of the statue, or its created
state, so to speak, may always transcend its form,
or its made state. For this reason you will go on
for ever reproducing your statue, or modifying its
actual form, the better to express its ideal being.
Now all this shews, analogically, that the crea-

ture, *quâ* a creature, or in so far as he is related
to God, is never absolute or independent, confessing to have once upon a time derived life from
his Creator, but to have possessed it ever since
in himself. On the contrary, he is a *perpetual*
creation, never ceasing at any moment to be
created, that is, to be instinct and vivid with the
life or personality of his Creator.

But now you will observe, that for this very
reason, the creature must be *made* as well as
created. Simply because there is no *essential*
discrimination or discrepancy between Creator and
creature, simply because, in other words, the
Creator constitutes the sole and total being of
the creature, it is evident that the latter can
never come to life or consciousness, can never
attain to the experience of his creation, unless he
undergo a *formative* process as well as a creative
one, and so become defined to his own intelligence. For as we have seen, making or formation
is always a finite process, is always a strictly
generative process, implying the development of
one thing out of another, and hence falling within
the range of the finite intelligence. Were the
creature simply created, he would be pure spirit,
undistinguishable from his Creator; he would be
without form or selfhood, hence unintelligible to
himself, or unconscious. For the created intelligence is incapable of apprehending truth in its
infinitude, or save as it is reflected in finite forms.

We can no longer doubt, then, that the creature must also be *made* or formed, that is, must also be outwardly projected and defined to his own recognition, as well as spiritually conceived or created. Otherwise that mythic and primeval night in which all cosmogony begins, would reign alone where now the human mind discloses the varied imagery of God's endless perfection. And to give this requisite external projection or definition to God's creature, is, as we have already seen, the precise and unswerving ministry of what we call Nature. Nature is simply a formative process, which separates the creature to his own intelligence from the Creator, by giving him finiteness, by making him self-conscious. Viewed in whole and in part, Nature is nothing else than the mould, or form, in which God's stupendous creation is run, in order to its becoming conscious of itself, and so finally conscious of God, its truer and better self. Creation is a purely spiritual and invisible process, known only to the creative mind. Our natural generation is only a needful mirror of that sublimer process, is only a necessary medium through which the creature thus created becomes pronounced to his own consciousness, and so qualified for the blissful destiny that awaits him, namely, ultimate conscious unity with infinite Goodness, Truth, and Power. In short, spiritual life alone is real or substantial; natural life is purely formal, and hence intensely subordinate to that. Nature yields us only the appearance or semblance of being to a limited intelligence. Being itself is supernatural, and therefore unlimited.

Thus the inspired cosmology involves in its construction a far deeper philosophy of life than has yet transpired in our best books of philosophy so called. It is, in fact, actually based upon a discrimination so sharp between object and subject, between substance and form, between Creator and creature, as would put our most accredited philosophy out of its wits even to conceive of it, and make it incontinently renounce creation as a sheer incredibility, if not impossibility. For the discrimination in question places a gulf between God and man, between Creator and creature, not less impassable than that which separates substance from shadow, or the reality of a thing from its merely mirrored semblance or appearance. Surely our best philosophy would grow frantic if it were required, on any such data, to construct an intelligent theory of creation. And this, simply because philosophy wholly overlooks Christianity as furnishing the sole competent logic of creation, or in other words, because philosophy completely ignores another all-pervading feature of revealed and historic truth, which I shall now proceed to bring before you. No truth is so intimately and characteristically blent as this with the living experience of man: no truth has been so *exclusively* operative (I might say) upon the developments of human history—all history, in fact, being only its protracted witness and reverberation : and yet our puffy and pretentious Philosophy, nevertheless, disdains even to name its name.

The truth in question, when viewed on its human side, is known under the familiar name of REGENE-RATION; when viewed on its Divine side it is called REDEMPTION. Let us first fix our attention upon it as it presents itself to us in the sacred symbol.

As I have just said, the opening page of Revelation sharply discriminates between being and form, between Creator and creature. We have first the Creator presented to us, the creature being still undeveloped. God creates or gives being *ab origine,* but the result of the creative energy is still invisible, the creature being unformed and unconscious, being *without form and void.* Creation is yet altogether spiritual, shut up to the creative mind, but the order of it as there existing is clearly given in the programme of the operations of "the Divine spirit, when brooding over the face of the waters." Here the ideal form and destiny of man are symbolically but vividly sketched under the various features of the six-days' work and the seventh-day rest, his constitutional divisions of affection and intellect, with all their minor derivations and powers being distributed under the names of heaven and earth, the greater and lesser lights, the day and night, the grass and herb and fruit-tree, the moving creature, the flying-fowl, the fish, the cattle and creeping thing, and so forth, until in the sixth or final working day, the dominant or binding form of all these passive and active powers is complete in man the image of God, male and female.

This is a picture of man on his created and unconscious side, or as he exists to the Divine mind exclusively, for we are immediately informed that such were the births of the heavens and the earth in the day when *they were created*, that is in the day when the Lord God made them, before they were visible to any one else, and when He made every plant of the field *before it was in the earth*, and every herb of the field *before it grew :* for as yet there had been no rain upon the earth, and no man existed. But now next of course we go on to read (always symbolically, remember) of his natural formation, and of his being placed in a garden, and his leading a life there of Paradisiacal ease and opulence. I say "next *of course*," because, as we have already seen, our natural selfhood, or the life we derive from nature, is the necessary basis of our spiritual development, is the indispensable *matrix* or mould by which we attain to our true self-consciousness as a purely Divine creation. Unless we had enjoyed this previous lower experience, unless we had first known ourselves naturally, or been indued with natural form, our higher

or spiritual and divinely-given selfhood would have
been destitute of all *measure,* of all *ratio,* would
have been without any *continent,* so to speak, by
means of which it could become manifested and
appropriated.  For example, if it were not for my
natural ties to my parents, to my brothers and
sisters, to my uncles and aunts and cousins, to my
neighbours and friends, to my fellow-countrymen
and my race, my selfhood, instead of becoming
enlarged to universal dimensions would remain
imprisoned under my physical limitations, and so
present no form adequate to the Divine inhabitation.  Thus these natural ties of kindred and of
race, which gradually universalize the form of my
consciousness, and endow me with a world-wide
selfhood, are merely the germ, are merely so
many rude husks and fostering envelopes, out of
which is born in fulness of time the consummate
and immortal spiritual flower.  These natural ties
*fix* as it were the supernatural reality, so enabling
me, the child of a day, to become woven upon the
substance of God, and breathe everlastingly the
atmosphere of His incorruption.  They are the
mirrored semblances of the eternal and invisible
Truth, and unless that truth had this preliminary
projection to our sensible experience, it would for-
ever remain impenetrable to our rational or spiritual understanding.

In Adam then, formed from the dust and
placed in Eden, we find man's natural evolution

distinctly symbolized—his purely instinctual and
passional condition—as winning and innocent as
infancy no doubt, but also, happily, quite as
evanescent. It is his purely genetic and *pre-moral*
state, a state of blissful infantile delight unper-
turbed as yet by those fierce storms of the intellect
which are soon to envelope and sweep it away, but
also unvisited by a single glimpse of that Divine
and halcyon calm of the heart, in which these hideous storms will finally rock themselves to sleep.
Nothing can indeed be more remote (except in pure
imagery) from distinctively *human* attributes, or
from the spontaneous life of man, than this sleek
and comely Adamic condition, provided it should
turn out an abiding one : because man in that
case would prove a mere dimpled nursling of the
skies, without ever rising into the slightest Divine
communion or fellowship, without ever realizing a
truly Divine manhood and dignity. He is still a
mere natural form sprung from the dust, vivified
by no Diviner breath than that of the nostrils,
mere unfermented dough, insipid and impracticable : and the Lord makes haste accordingly to
add the spiritual leaven which shall ensure his
endless rise into human, and ultimately Divine
proportions. He brings him Eve, or *spiritually
quickens him ;* for Eve, according to Swedenborg,
symbolizes the Divinely vivified selfhood of man.
The Adamic dough, heavy and disheartening before, becomes lively enough now in all conscience,

becomes instinct and leaping with vitality, although that vitality has no more positive form
than a protest against death, a struggle against
mortality. Thus had we had Adam, "male and
female," alone for a progenitor, we should never
have emerged from our Edenic or infantine gris-
tle : we should have remained for ever in a state of
Paradisiac childishness and imbecility : in a word,
we should have been destitute of our most human
characteristic, which is history or progress. We
should have had mineral body and consequent
inertia, no doubt: we should have had vegetable
form and consequent growth ; we should have had
animal life and consequent motion : but we should
have been without all power of human action,
because we should have lacked that constant per-
meation and interpenetration of our spirits by the
living spirit of God, which weaves our pallid natural annals into the purple tissue of history, and
separates man from nature by all the plenitude
and power of incarnate Deity. Human history
dates from Eve. Existence dates from Adam, but
life, or progress towards God, begins with Eve :
hence she is named Eve, mother of all living. It
is Eve, or the vivification of our natural earth by
the Divine spirit, which disenchants us of our
long Adamic babyhood, which emancipates us from
Eden, which shews us first how full of inward
death and horror is that imbecile being we have
in Adam, only that we may subsequently see into

G

what pregnant and delicious life this death becomes transmuted by God. For now begins the moral experience of man, that purely *transitional* stage of human experience, in which man discovers the corruption and death he has in himself as only naturally vivified, or unvivified by God, and which separates him on the one hand from the merely animal life of instinct, and on the other from the truly human one of spontaneity. This is the sole force and function of our moral experience, to release us inwardly from the Adamic clutch, and so leave us free to the Divine indwelling. God has no conceivable quarrel with the Adamic life in itself, but only as claiming our spiritual allegiance. On the contrary, when the natural man spontaneously disposes himself to serve the spiritual, he will find, unless all prophecy is illusory, a far ampler satisfaction of his wants Divinely secured to him, than he now so much as dreams of.

But I shall be obliged to reserve what I have still to say, for another Letter, and am meanwhile

Yours truly,

————.

# LETTER XVI.

*Paris, Jan.* 1, 1857.

MY DEAR W.

WE saw at the close of the last Letter, that the literal form which the revealed Truth puts on, in order to adapt itself to the carnal apprehension, is that of a Divine regeneration of man. The sacred narrative represents the Creator as developing a new life from out the ribs of the natural Adam, or as giving His creature a spiritual new-birth. The natural man is first represented as standing at the head of all created things, or rather, as involving in himself all lower forms of life: (for by Adam naming all cattle, *etc.*, is signified that all earthly things derive their quality from man, being all only so many fragmentary exhibitions of human nature, *name* in the spiritual world meaning quality or character): and then we are told that God finds him still insufficient to himself, or *alone*, and proposes to furnish him with a suitable companion and helper.

What is the force of the word "alone" in this passage? What precise infirmity of the natural

G 2

man is indicated by it? We may be very sure
that a superb significance attaches to all this
symbolism if we can discover it, and with the clue
we already possess, I think we have no reason to
distrust our ability to do so. Let me say, then,
without any more circumlocution, that by Adam,
or the natural man, being *alone*, is meant, that
man on all his *constitutional* side, or in so far as
he is related to nature and society, is destitute of
real freedom, is without any true selfhood, or
possesses only a phenomenal life, and hence is
subject to mortality. Let me make this plain.

If you have read Swedenborg's luminous exposition of the laws of spiritual existence with
due attention, you will have learned that all our
natural affection and intelligence is derived to us
from spiritual association. I have no self-love nor
brotherly love, no love to my own body nor to
the world, no love to parent or child, to brother
or sister, to friend or neighbour, to man or
woman, which is not a strict inflow to my heart
from spiritual societies, celestial and infernal, with
which I am connected by my natural generation,
that is, by all my past ancestry. Neither have
I any natural intelligence, any sense of good and
evil, true and false, sweet and bitter, hard and
soft, light and dark, but what comes to me from
similar association. In a word, my nature is a
simple inheritance or derivation to me from my
past ancestry: it is nothing more and nothing

less than an aggregate image and reflection of all
the so-called good and evil men and women, to
whom in endless complication I owe my sensible
production. Thus to all the extent of my purely
*constitutional* limits, that is, to all the extent of
my physical and moral nature, I am a mere
helpless product of the invisible spiritual world,
without human character or dignity. Indeed, I
am so wholly constituted as to my affection and
thought by spiritual association and influx, that
I should even be without my distinctive human
form, if it were not for God's profounder grasp
of me, or the quickening operations of His infinite spirit within our nature. Had I had no
profounder life than that which binds me to
nature and society, were I not related to God
more profoundly than I am related either to my
own body or to my fellow-man, I should have
remained mere dove or serpent, mere horse or
lion, mere sheep or tiger, to the end of the chapter: that is, I should have remained just what
my spiritual association made me, a living animal
without true freedom or selfhood, without any
Divine quickening, and consequently without any
power to rise above the lot of my nature.

This is what the good book pronounces a *lone-
some* condition, in which obviously it is not good
for man to continue, because while it lasts he is,
though apparently conjoined with God, in reality
disjoined. For the human mind is here presented

to us as yet only in a celestial condition, that is, in its infantile beginnings; and although this condition is one of seeming innocence, yet the least reflection shews that the innocence is more apparent than real, more superficial than solid, being of that sort which characterizes lambs and doves rather than humanity. It is destitute of the human element, which is spontaneity, and hence will not keep. It is the innocence which flows into us from spiritual association, which we inherit from our past ancestry, and is consequently incapable of constituting our true individuality. As to my natural or inherited genius, I may be as guileless and harmless as all the lambs and doves extant; yet nothing shall hinder me spiritually or individually becoming perhaps as ravenous as the wolf, as cunning as the fox, as lordly as the lion, as venomous as the serpent, simply because this guilelessness and harmlessness have no root in my proper spontaneity, that is in my God-given self-hood, but are reflected upon me from chance spiritual association. Swedenborg was never able to discover any angelic existence which was spontaneously good, or good of itself. On the contrary, he found that the highest angel, when dissociated from his fellows, and brought into contact with lower influences, became as lascivious and vile as any devil. In short, he discovered that the angelic goodness was invariably contingent upon harmonic association, or depended upon a rigorous previous elimination of evil; and hence was anything but spontaneous. Evidently, then, this celestial goodness would be a very poor rest for the Divine creation. Creation would in that case be like an air-built house without any foundation in the earth. And yet our Adamic side, our merely natural selfhood, our *constitutional* life, so to speak, has no profounder source than this. Man is naturally only what he is made by spiritual association. Every affection he feels, every thought he experiences, every breath he draws in fact, is an influx from spiritual companies in which all unconsciously to himself he has been immersed from birth: and consequently if he were wise, he would, as Swedenborg says, appropriate neither his good nor his evil to himself, but dwell incessantly in a region of Divine life and peace, undisturbed by that mean conflict. But this wisdom comes in its own sure time. We will not ask the child to anticipate the man, lest the man himself be spoiled. Adam is but the celestial infant unweaned as yet from the maternal bosom, or without real selfhood: let us wait for the second and sublimer Adam, for the adult and ripened manhood of the race, to see this selfhood triumphantly asserted. *Handle me and see,* said the Christ, *for a spirit* HAS NOT FLESH AND BONES *as ye see me have.* In symbolic literature, "flesh and bones" signify the natural life or selfhood, what Swedenborg calls the external man. Angels and spi-

rits are thinly clad in this particular, because,
inasmuch as their life is a perpetual derivation or
influx to them from parent societies, they have
manifestly no independent selfhood, no life in
themselves. They have only the appearance of
such life, never the reality. They do good, says
Swedenborg, only AS OF themselves, never really
OF themselves. For the creature is really, by the
very terms of the proposition, without selfhood,
except what he derives from the Lord. Every
thing short of this is a mere flatulent fallacy, " a
mere dead nothing," says Swedenborg, " though
it seem to us so real and important, yea, our very
all." " Natural wisdom," he says elsewhere,
" laughs when it is told that man has no selfhood,
his selfhood being only a fallacious semblance:
and it laughs still harder when it is told, that the
more man believes in such apparent possession,
the less selfhood he really has, while the angels
on the other hand, who don't believe in it at all,
and reject it from them, are filled with a most real
selfhood from the Lord."—*A. C.*, 2654. See also
the *Divine Providence*, 308-9. Thus even the show
or semblance of life we have by nature in ourselves, attributes itself to the DIVINE NATURAL
man. It is only because the Divine Love is capable of eventually endowing us with real freedom, with real selfhood, that this seeming freedom
or selfhood becomes previously practicable. In a
word, if we had not been ultimately destined for a

spontaneous life or righteousness, that is to say, for a life and righteousness which shall inhere in ourselves and not be derived to us from without, we should never have exhibited a symptom either of moral or of physical consciousness.

Understand, then, that Adam symbolizes the celestial or rudimentary condition of humanity, the state of infantile innocence and ignorance we are in before the dawn of morality, before we have eaten of the tree of knowledge of good and evil. When Swedenborg conversed with persons of this transparent type, with angels corresponding to this germinal and tenderly beautiful aspect of humanity, he found them of an exquisitely innocent and docile deportment, attributing neither good nor evil to themselves, and enjoying ineffable peace and felicity in the Lord. But the tendency to selfhood, or the inappeasable desire to be wise from themselves, must have been latent in all their specific genius, since it came out so fully in that of their descendants. These descendants, says Swedenborg, were unwilling to be led of the Lord, or would be wise from themselves: and though they had no conception as we now have of the Divine glorification of human nature, and consequently incurred the condemnation of an unenlightened conscience, still their aspiration was intensely human, betokening more than all things beside the presence and vivacity of the Divine spirit within them; and accordingly selfhood was granted them vivified with all Divine love and wisdom. *And the rib which the Lord God had taken from man, made he a woman, and brought her to the man. And Adam said, This is now* BONE OF MY BONES AND FLESH OF MY FLESH,—*therefore shall a man leave his father and his mother, and shall cleave unto his wife, and they shall be one flesh.* To leave one's father and mother, means spiritually, to cease being an internal man, that is, to cease being dependent upon celestial and spiritual influence. And to cleave to one's wife means spiritually to become enlightened and enlivened from oneself, that is, by the operation of the Divine natural humanity. In short, the whole passage means our ceasing to vegetate and beginning to live: it indicates the transition from a merely constitutional and imbecile existence into a life of plenary Divine contentment and power. The 'woman's being called by the man "bone of bones and flesh of flesh," signifies the intimate dearness and nearness of the selfhood to the human heart, and recalls those profound words of Christ already quoted, *a spirit hath not* FLESH AND BONES *as ye see me have.* It is as if He said in other words, "a spirit is utterly void of natural force, or life in himself, and yet I possess this life even after my death in undiminished vigour." This *flesh and bones,* this divinely vivified natural selfhood, is what we have all been struggling for from the beginning, is what we are all now miserably prey-

ing upon ourselves and upon each other for the
lack of. What a profound though all unconscious
confession of the fact, broke from the bosom of
my manly friend Thackeray the other day, when,
in one of his lectures on the four Georges, he thus
painted the insanity of George the Third. I
quote from a newspaper :—

"'History,'—thus concluded the lecturer, amidst
the solemn silence of the audience,—'presents no
sadder picture than that old man, blind and de-
prived of reason, wandering through his palace,
haranguing imaginary parliaments and reviewing
ghostly troops. He became utterly deaf too. All
sight, all reason, all sound of human voices, all
the pleasures of this world of God, were taken
from him. Some slight lucid moments he had,
in one of which the queen, desiring to see him,
entered the room and found him singing a hymn
and accompanying himself on the harpsichord;
when finished, he kneeled down and prayed aloud
for her and for his family, and then for the nation,
concluding with a prayer for himself that God
would avert his heavy calamity from him; but if
not, that He would give him resignation to submit
to it. He then burst into tears, and his reason
again fled. *What preacher need moralize on this
story? What words, save the simplest, are requisite to tell it? It is too terrible for tears. The
thought of such misery smites me down in submission before the Ruler of kings and men—the Mo-*

*narch supreme over empires and republics—the inscrutable Dispenser of life, death, happiness, victory.* Oh, brothers, I said to those who heard me first in America—oh, brothers, speaking the same dear mother-tongue; oh, comrades, enemies no more, let us take a mournful hand together as we stand by this royal corpse, and call a truce to battle. Low he lies to whom the proudest used to kneel once, and who was cast lower than the poorest, whom millions prayed over in vain. Driven off his throne, buffeted by rude hands, with his children in revolt, the darling of his old age killed before him, old Lear hangs over her breathless lips, and calls—Cordelia, Cordelia, stay a little.

> Vex not his ghost, O! let him pass, he hates him
> That would upon the rack of this rough world
> Stretch him out longer.

Hush strife and quarrel over the solemn grave! Sound, trumpets, a mournful march. Fall, dark curtain, upon his pageant, his pride, his griefs, his awful tragedy!' "

"The thought of such misery," says this sincere and tender soul, "smites me to the dust before the awful Ruler of kings and men." What a melting cry of anguish is here! This masterly writer, who sounds at will all the depths of human nature, is no stronger nor wiser at bottom than the rest of us: he too feels life insecure, he too lifts a pallid face in prayer to God lest some hideous calamity engulf his fairest hopes. Few persons have maintained their natural *naiveté* and candour so unbronzed by contact with the world, as this great and hearty Thackeray, this huge, yet child-like, man; but if the secret bosom of men were canvassed, there would be none found who does not profoundly sympathize with him. We are all of us without real selfhood, without the selfhood which comes from God alone. We have only the shewy and fallacious one which inflows from the spiritual world, and which is wholly inadequate to guarantee us against calamity. We shiver in every breeze, and stand aghast at every cloud that passes over the sun. When our worthless ships (which we ought to be ashamed of building, which we ought in fact to hang our shipmasters for building) go down at sea, what shrieks we hear from blanched and frenzied lips peopling the melancholy main, perturbing the sombre and sympathetic air, for months afterwards! When our children die, and take back to heaven the brimming innocence which our corrupt manhood feels no use for, and therefore knows not how to shelter; when our friends drop off; when our property exhales; when our reason totters on its throne, and menaces us with a downfall; who then is strong? Who, in fact, if he were left in these cases for a moment to himself, that is, if he were not steadied in his own despite by the mere life of routine and tradition,

but would be ready to renounce God and perish?
So too our *ennui* and prevalent disgust of life,
which lead so many suffering souls every year to
suicide, which drive so many tender and yearning
and angel-freighted natures to drink, to gambling,
to fierce and ruinous excess of all sorts: what are
these things but the tacit avowal (audible enough
however, to God!) that we are nothing at all and
vanity, that we are absolutely without help in ourselves, and that we can never be blessed and tran-
quil until God take compassion on us, and conjoin
us livingly and immortally with Himself?

This be assured, my friend, is the inmost meaning of human history. To become conjoined with
God naturally as well as spiritually—this is the
great destiny of man which we are at last on the
very verge of realizing, unless some new and sick-
ening imbecility, some new inrush of merely fallacious and sentimental life, set us back again in
the direction of the base earthly Adam. I confess I have my fears for a portion of the race
when I read of the insanities and inanities of our
modern ghost-mongers, American and English.
Since the beginning of history the Divine natural
humanity has been doing its best to struggle into
conscious life: that is to say, *a divinely perfect
order for the natural mind,* for man in nature, *has
been persistently seeking to come into clear scientific speech and recognition.* What has hindered
it doing so? Nothing whatever but the overbear-

ing *prestige* of the so-called spiritual world : nothing whatever but the remorseless tyranny exerted thence over the natural imagination of the race. The influx of that world into nature has always so inflamed our merely natural affections, has bound us so helplessly to our parents and grannies, to our uncles and cousins, has kept up such a perpetual vivification, in other words, of the most narrow and abject natural prejudice in the mind of the race, that human progress has been almost impossible, and would have been quite so, if God had not mercifully limited the sway of priesthoods, or rather systematically deflected their influence to the cordial fomentation of our distinctively secular ambitions and aspirations. As I have said before, we live under the Iscariot dispensation, for Judas was one of the chosen twelve, and has his inalienable significance in the history of the Christian Church. Thus we find that the inevitable progress of the Church itself has utterly sapped our reverence for the spiritual world, or left it a mere empty tradition : we find that the mass of mankind in Christendom has, *through the increasing worldly pomp and affluence of the Church itself*, become *disinterested*, so to speak, in the spiritual world, and are turning themselves with boundless goodwill on every hand to ask rather, what are God's marvels of order and wisdom, of love and mercy, for this despised and neglected natural world. In one word, men are getting thoroughly tired of their old, seeming, merely constitutional and finite life, and with panting sides yearn and pray for one fragrant breath at last of their real, Divine, and infinite one. And the consequence is, that the Divine Love is at length beginning to avouch its redeeming presence in the earth, beginning to glorify the common natural life, beginning to shew the indiscriminate human selfhood aglow with invention, with skill, with power, with grace, with every Divine faculty in short, and so to commend every man to his brother's unlimited respect and benediction as a temple instinct with Divinity.

In this critical condition of things, when every mountain-top blushes with the splendour of oncoming and incarnate Deity, a set of belated Rip Van Winkles, who have done nothing but snore while others were astir for long centuries, suddenly make the discovery (which no wide-awake person has ever needed to make, so cruelly hampered and oppressed has his proper human force always been by his overwhelming consciousness of the fact) *that the spiritual world exists, and exerts an intimate and enormous influence upon nature.* Prodigious! Such an opportune discovery too, betraying the very heyday and sabbath of drowsiness, to take place just as the enlightened bulk of Christendom are ripe for the conviction, that nature has her harmonies no less Divine than those of spirit, and that if the soul has hitherto

claimed the blind obedience of the body, it has
only been because both alike have imperfectly
realized their true destiny, or failed to subserve
that Divine and perfect life in man, to whose commanding needs they are quite equally subordinate.
And after all our sceptical discoverers seem by no
means sure of their discovery.  The way they run
after additional evidence, the fervour with which
they receive every reiterated joggle of the ma-
hogany, illustrated erewhile by spirits more ardent,
but incalculably less mischievous ;* the glee, in

* " It is believed by many," says Swedenborg, "that man may
be taught of the Lord by spirits speaking with him : but they who
believe and wish this, are not aware that it is connected with
danger to their souls.  So long as man is in the world he is indeed
in the midst of spirits, as to his spirit, but these spirits do not know
that they are with man, any more than man knows that he is with
them.  The reason of this ignorance is, that their *immediate* conjunction takes place in their affection, whereas they are only *me-*
*diately* conjoined in the sphere of thought, natural thought having
only a *correspondential* relation to spiritual thought, and relationship by correspondence leaves one party completely ignorant of the
other [makes one, in fact, the inversion of the other].  But whenever spirits begin to speak with man, they come out of their own
spiritual state into man's natural state [that is, out of a correspondential into an actual relation], and then they know that they are
with him, and conjoin themselves with *the thoughts which flow*
*from his affection, and from those thoughts converse with him.*
They cannot enter into anything else, without at once separating
themselves from man, for the law of spiritual intercourse is, that
similar affection conjoins people, and dissimilar disjoins them.
Hence the speaking spirit is *necessarily in the same principles with*
*the man to whom he speaks, whether such principles be true or*
*false*, and hence he is sure to excite those principles, and by the

short, with which they hail every trivial proof of
a haunted side to our baser nature, of an under-
hand and sneaking ghostly interference permitted
through the crevices and rat-holes of our still
most disorderly natural and associated existence:
all this shews, I say, that they are even yet in-

added force of his will strongly to confirm them. From all this it
is clear that a man can never speak with, or be otherwise operated
upon by, spirits essentially different from himself, so that enthu-
siasts always come in contact with enthusiastic spirits, fanatics
with fanatical ones, heretics of every complexion with heretical
ones, and so forth. All spirits speaking with man are precisely
what they were when in the world, as I have known by multiplied
experiences. And what is ridiculous is, that when a man believes
that the Holy Spirit speaks with, or otherwise operates upon, him,
the spirit is in the same persuasion and fancies himself the veritable Holy Ghost. This is common with enthusiastic spirits. From
these considerations it is evident how dangerous it is to court spiritual intercourse. Man does not know the character of his natural affection, whether it be good or evil, or whether it be conjoined with heaven or hell; and if he be at all conceited of the
intelligence which flows from that affection, his familiar spirit will
be sure by fanning that intelligence to confirm his particular affection, and so possibly plunge him into irremediable disaster. The
Pythonics, &c., &c., were formerly of this sort: but the children
of Israel, or the representative church of the Lord, were forbidden
to seek spiritual communications under penalty of death." See
*The Athanasian Creed*, 74. See also *The Divine Providence*, 321.
" *Those who are instructed by influx as to their beliefs or conduct,
are neither instructed by the Lord or any angel, but by some fana-
natical spirit or other, and are seduced. All really Divine influx
takes place by an enlargement of the understanding, growing out
of an enlarging love of truth.*" I recommend these words to the
serious regard of every one whose mental tabernacle has begun to
be disquieted by rats. See *Appendix A.*

completely assured, and regard spiritual existence much less in the light of a truth than of a probability. Who can say that minds of this cast will ever be satisfied, even when they go into what they call the "*spirit-world:*" who can say that even then they will not go about for further evidence and testimony, and, like the sieve of the Danaides, never know when they shall have got enough?*

But however all this may be, I want you distinctly to observe that the spiritual world is utterly void of claim to our rational regard, except as ministering to our exclusively *finite* side, to our purely *constitutional* endowments, as distinguished from our proper life. It has no direct relation to our life, but only an indirect one through our physical and moral natures, through our natural and social existence. My life is spontaneous and free, flowing from the immediate presence of God in me. No doubt this life demands as a platform, or basis of its own manifestation, my physical and moral existence, just as the upper stories of a house demand an underground foundation: but life is no more to be confounded with its mere subter-

* I count several beloved and admired friends in this movement, who predict excellent results from it. While I rejoice that their own ample and powerful wills shield them individually from the mischiefs which inhere in an undue familiarity with these ghostly Jeremy Diddlers, these spiritual ticket-of-leave men, I all the more abhor and deplore the frequent and fearful disasters which ensue to feebler organizations.

raneous conditions, or with existence, than the
drawing-rooms and bed-chambers of a palace are
to be confounded with its kitchen and larder. All
the wretchedness of our past and present history
refers itself in some shape to this shallow and
pestilent error. The cleanly and beautiful temple
of God in our souls is incessantly overrun with
spiritual vermin in consequence; we are daily
chased from corner to cupboard, from cellar to
garret, by stenches so infernal as to put us to our
wits' end for a remedy: and no remedy appears
but at once and manfully to learn to separate between existence and life, or what is the same thing,
to compel the spiritual world equally with the natural one into the humble harness of use, into the
undeviating and eternal subjection of God's life in
man. Thus if any nasty spiritual person should
contrive to come to us through the reeking chinks
of our still unscientific mental sewerage, saying
that he has been divinely relegated to a certain
charge over our servile and constitutional interests,
over our natural affections and intelligence, let us
tell him in return that he is a very precious ass to
affect a mission of that nature, since all the good
we do each other in such connection is strictly
contingent upon our being utterly unconscious of
it. And if he go on, on the other hand, to allege
that he bears any the faintest conceivable relation
to our real and immortal parts, or to that life
which is alone worth our thought because it alone

comes from God, let us greet him with a cachin-
nation so hearty and derisive, as shall bid him
instantly disperse, nor ever shew his foolish face
again within the breezy realm of cockcrow.  Understand well that no human being, angelic or
diabolic, touches us except circumferentially :
never in the regal and transcendent plane of Life,
but only in the servile plane of Law.  In so far
as I am a fixity, that is to all the extent of my
relations to nature and society, of my physical
and moral existence, I am intimately dependent
upon angel and devil.  I have neither health of
body nor sanity of soul but by a preponderant
influx from heaven; nor have I disease of body
and insanity of soul but by a preponderant influx
from hell.  Thus if I had no commanding life in
God, I should be the mere chance puppet of these
warring influences, and go on myself to swell the
ranks of angel or devil to all eternity, as my own
inherited tastes might decide.  What I feel bound
then by my supreme loyalty to the Divine life to
do, is, to shake my cordial fist at both angel and
devil, bidding one and the other alike to observe
a respectful distance.  I will have no private relations with either of them.  If between them they
can contrive any benefit to my *common* nature,
physical and moral : if by the growing subordination of hell to heaven, and of heaven to the
Divine, the ordinary level of our natural and
social existence becomes elevated, I no doubt, like

every body else, will prove a grateful participant of that boon: but I will accept no special advantage from either quarter. In fact, I would not give one fig to call all the good that gladdens any heaven my own; nor would it cost me one pang of self-reproach to find myself charged with all the evil that festers in any hell: simply because I am profoundly sure that in both cases alike the possession would be only apparent, not real; that is to say, would attach to me exclusively on the side of my moral or *quasi* freedom, and not on that of my spontaneous and genuine freedom.*

But it is high time that we got back to our great symbolic starting-point and progenitor— Adam. But as I shall still have much to say in this connection, I shall probably consult your convenience by deferring it to another letter.

Yours truly,

————.

* See *Appendix B.*

# LETTER XVII.

*Paris, Jan. 5th,* 1857.

My DEAR W.,

WE have seen that the truth which inspires all
revelation and enlivens all history, is the truth
of the Divine vivification of human nature, or of
God's essential humanity. God gives life, no
doubt, to angels and spirits, but only because
angels and spirits are partakers of human nature,
because they are germinal or rudimentary men.
In short, A TRULY INFINITE GOODNESS AND WISDOM
INFORM AND ANIMATE HUMAN NATURE, AND THAT
NATURE ALONE. This truth which only our tardy
docility in Divine things, in other words, our infatuated self-conceit, hinders us seeing, is the sole
interior meaning of revelation, constitutes the
entire spiritual burden of the literal dogma of
Christ's glorification or divinity. It is a truth so
utterly remote from the unassisted reason of the
race, that its clearest and most emphatic enuncia-
tion in the Christ, has been incessantly perverted
to the damage and degradation of our common

humanity. Men have been prompt enough to believe in God as the friend of certain distinguished persons, of certain regenerated or angelic specimens of the race : but it has never been credited that the Divine favour turned in every such case upon the fact, that the persons in question were only more truly men than others, were *more*, and not *less*, finished specimens of unalloyed manhood. We have always cheerfully said and sung : " Yes, we shall doubtless enjoy a Divine beatitude after life's fitful fever is over, that is, after we too shall have become angels; but so long as we remain mere men, we have nothing to expect but indifference at the Divine hands."

Now to what cause are we to attribute this inveterate ignorance and stupidity on our part ? Why obviously to the fact that we are still in our human babyhood, that our characteristic human life—the life we derive exclusively from God—is still almost unbegun, that creation in short is yet unachieved. *Our distinctive human life is really an immortal life, is so veritably grand and august as to place its true beginning only where all other things find their ending, namely, in death.* It converts death into its own immortal pasturage, turns it into its own prolific and exhaustless womb. For our distinctively human or characteristic life begins, only when the animal and moral life ends, only when our relations to nature and society, to our own body and to our fellow-man, have been

reduced to the *régime* of law. It is only because
God is my inmost life, because my proper human
force dates from Him instead of being inherent in
my physical or social conditions, that I am capable
of what no animal is capable of, namely, of con
science or moral power, which is the power of
transcending my physical and social constitution,
or of reducing nature and society to the service
of my individual tastes and attractions. Conscience attests the faculty which man alone possesses of separating himself from his merely finite
and constitutional environment, and allying himself with infinite goodness and truth. Every one
who has ever experienced a genuine moral afflatus,
every one whose conscience has not undergone a
hopeless pharisaic twist and sophistication from
the existing priestly corruptions of the Divine
name, knows that conscience is an invariable minister of death, is a perpetual flaming-sword turning every way to guard the Tree of Life, and will
therefore be quite ready to allow that the Divine
power alone is competent to sustain him under it.
Nothing whatever explains my moral experience,
or the operation of conscience in me, but the fact
that my life derives immediately from God. He
must be intimately present and busy in every distinctively human — that is, really individual —
breath I draw, in order to account for my surviving even for a moment the legitimate operation of
conscience. For conscience, freely operative, fills

H

my bosom with the pungent and stifling odour of
mortality, with the intimate and overwhelming
presence of death ; and nothing hinders this death
becoming instantly actual as well as sentimental,
outward as well as inward, natural as well as spiritual, but a conservative power within my nature
deeper than *myself*, but a living presence within
me infinitely more Divine than my present consciousness is ever prepared to ratify.   Our life is
always deeper than we know, is always more Divine than it seems, and hence we are able to sur-
vive degradations and despairs which otherwise
must have engulfed us.   Why does the animal
exhibit an utter dearth of conscience ?   Why, for
example, does he feel no inward monition of death
when he robs his fellow of a savory carcass ?   Because the animal's life is at best but a process of
dying, or is essentially mortal : because, in a word,
he is individualized only by his nature, and hence
is insensible to every motive but those of a mere
natural communism, which leaves him empty of
all Divine privacy or sanctity.   I, on the other
hand, being man, am essentially immortal, being
quickened or individualized by my nature only in
appearance, while in reality I am supernaturally
quickened.   It is the distinction of man to be
individualized by God alone, and hence to realize
true being only in dying to his seeming or spurious
one.   Accordingly when I rob *my* fellow of *his*
bone, or do him any species of injustice, I am

not like the animal, at peace with myself, but am
filled on the contrary with a poignant interior
anguish which flows down and poisons every spring
of natural delight.    It saps my most robust life
with instant decay, it smites my most ample and
clear-shining day into niggard and appalling night.
And this simply because a really infinite love and
wisdom embed human nature, quicken the human
form, and hence qualify me as they qualify no
animal for a life commensurate with all Divine
perfection.    Conscience is the negative attestation
of this truth.    It is the unripe aspect of the Divine
life in humanity ; it marks the period of spiritual
pregnancy or gestation in us, before the Divine
seed has taken appreciable form or come to self-
consciousness, and is attended consequently by all
those signs of spiritual nausea, distress and anguish, which announce in every sphere the descent
of new life.

      This, then, is the reason of our prevalent ig-
norance and stupidity in Divine things.    It is
because we are still uncreated, so to speak, because
we are still but *quasi* or prospective men in place
of real and consummate ones.    Our proper life is
essentially immortal because it comes from God,
but of course it cannot come to consciousness in
us so long as we remain spiritually subject to mortality, so long as our bosoms are the abode of all
unmanly cowardice and fear, or bring forth only
envy, malice, hatred, and every other fruit of

death. We are still too generally the abject
slaves of nature and convention to recognize our
proper human worth, and until we do this we are
of course ashamed to affiliate ourselves to the immaculate Goodness. God is the father of freemen
not of slaves, and therefore an instinct full of
worship keeps us still in cordial unbelief of the
Divine name, nor need we expect this unbelief to
be softened by anything short of the scientific
conception of human destiny, or the oncoming of
the Divine NATURAL humanity.

But if all this be true : if to be man be truly to
image God and realize all Divine blessedness : you
will reasonably inquire of me why the Scriptures
are not constructed upon the scientific acknowledgment of the fact? If to be man be all that
the creative Love and Wisdom desires in its creature, how comes it, you will ask, that the Scriptures first describe an apparent creation of man,
then represent him as falling from that condition,
and subsequently proceed to insist upon his regeneration or recreation? If to be man be all that
God requires in order eternally to bless us, why
do the Scriptures of truth exhibit man in this
divided aspect, under the lineaments of a first and
second Adam, or as the subject of an old and a
new birth? In short, why should the Divine
creation imply to the creature's consciousness his
own regeneration? The answer is not difficult.

You remember that in one of my earlier Let-

ters, the sixth, or seventh perhaps, if I am not
mistaken, I shewed you that revelation exacted
both a body and a soul, both a fixed or finite
earth and a free boundless heaven. An undefined
revelation, since it would be incognizable to the
finite or created intelligence, would be no revelation. It would be, like a disembodied soul, an
essential absurdity or contradiction. Every revelation addressed by God to the human understanding, must fall within the forms of that
understanding, under penalty of defeating itself.
Should the understanding be still sensuous and
infantile, revelation must clothe itself in strictly
corresponding forms, content to reserve its spiritual splendours for the maturer manhood of the
race, for those advanced periods of history when
reason having become emancipated from the burly
and deafening pedagogy of sense, shall reflect the
direct voice of God. All this is obvious enough.
But the true explanation lies deeper. The true
reason for the necessity which revelation is under
to obey the laws of the human form, lies in the
fact that the Divine life in man is a spontaneous
life, is a life in which man being directly vivified
by God shall do good of himself, *sua sponte*, of
his own accord, and no longer by self-denial.
Such being the case, it is obvious that God must
guard the rude and tender beginnings of selfhood
or freedom in man with exquisite jealousy, as one
guards the apple of his eye, because if these germs

should be blighted, or in any way prematurely
forced, our true freedom or selfhood would never
be realized, and our spontaneous or perfected life
accordingly perish before reaching its maturity.
Hence the Divine Love dreads nothing so much
as the suggesting a suspicion to man that his life
is outwardly derived to him, or does not inhere in
himself, and it never breathes a whisper accordingly but in tones most consonant to the creature's
consciousness.  It pays the most sedulous defer-
ence to the limits of the finite intelligence, and
never reveals itself but under the most rigorously
human lineaments.  Should the creature, for example, misled by his senses, cherish an erroneous
notion of himself, should he instinctively esteem
himself the parent source of his own good and
evil, and habitually assume therefore the responsibility of his own actions, the letter of revelation
must perfectly authenticate this fallacious instinct,
nor utter the slightest syllable in derogation of it,
under penalty of defeating its own aims.  The
aim of the Divine Love is to develope in its creature a spontaneous force, a force flowing from a
perfect conjunction of infinite and finite, or internal and external; and manifestly the only possible nucleus of this force is the unviolated natural
instinct of the creature.  Our natural instinct of
freedom or selfhood is but the germ or egg of the
fully perfected Divine and spontaneous life, and
hence unless it were most tenderly regarded by

God, unless its rudest physical beginnings and its subsequent social or moral enlargement, were most zealously fostered and cultivated by God, even as a judicious gardener fosters and cultivates the roots and stem of a plant, it would be destroyed, and with it all the Divine and immortal promise with which it is big.

From the absolute necessity of the case, then, the letter of Divine revelation, or its bodily form, exactly reflects the shape of the finite intelligence. It is never the scientific expression, but only and at best the dim memorial, the remote symbol or picture, of its own interior spirit. It does not express the Divine life in man as that life exists to its own consciousness, but only as it exists unconsciously, or to the eyes of an infirm intelligence. Your image in the looking-glass is not the living expression of yourself, but only its lifeless effigy addressed to an outward eye, the eye of sense. It is not a reflection of your real *life*, but only of your phenomenal *existence*. It does not express your conscious and invisible self, but only your unconscious and visible one : not that which really *is*, but that which *appears* to outward eyes. Your conscious or real self stands expressed only in your action, in your work, in what you freely effect. It is only your unreal and apparent self which reveals itself in the fleeting image impressed upon the glass. So precisely the Divine life in man finds its living or conscious expression, only

in the spontaneous life of man, only in the free
play of his taste, of his productive energies, of
those marvellous æsthetic aptitudes which constitute what we call *genius*, or which obey the law
of spiritual attraction and disown every outward
law. It is only its *seeming* character, the aspect
it bears to a sensuous intelligence, an intelligence
inferior to itself, which stands revealed in the
visible symbol. We must not only not be surprised therefore, we must intelligently expect, to
find the letter of the Divine revelation in flagrant
disagreement with its spiritual contents. This
disagreement is the conclusive attestation of its
reality as a literal Divine revelation. If it had
stooped incidentally to discharge any of those
pedantic offices upon which our modern critics
suspend their acknowledgment of its divinity: if
it had undertaken, for example, incidentally to
rectify our natural prejudices about the solar system, about the deposit of dew, about the possibility of miracles, and so forth: it might certainly
have thus anticipated the growth of the scientific
understanding in man, but to have anticipated
that development would have been to defeat
it. The scientific evolution of the human mind
marks its interior expansion, its growing spiritual
emancipation from the empire of mere natural
prejudice, from the dominion of established authority, routine, or custom. Science, in short, is
the Divinely-perfect body of the Divinely-perfect

mind of the race: the fixed and indestructible
earth at length upon which God's new and permanent heavens are adequately based. If therefore God had thrust any of its truths upon the
mind of the race prematurely, or before its advancing interior expansion naturally clothed it
with that summer foliage, He would have acted
like a silly gardener who in the height of winter
decorates his lifeless trees with artificial leaves,
and so insults the instinct of truth and fitness in
every genuine bosom. If the tree were capable
of estimating the husbandman's antics, its life
would scarcely revive: for it would say, "This
silly man is just as content with seeming fruit as
real; he wants not that I should really bear fruit,
but only appear to bear it, and hence it is all the
same to him whether I am alive or dead." But
man is capable of estimating *his* husbandman's
ways, and if therefore he saw God decorating him
either with leaves or fruit in anticipation of his
natural powers, or which his own intellect and
will had not honestly engendered, his profoundest
instincts of freedom or selfhood would be instantly
undermined, and creation declare itself the empty
farce it really was. The scientific mind of the
race, in short, is the DIVINE NATURAL mind, is
the slow accretion of its interior celestial and spiritual experience, of its varied life of affection and
thought; and hence it constitutes the very last
result of human history, the crowning achieve-

ment of the creative Wisdom.   Obviously then it admits only of a symbolic anticipation, or a revelation by means of natural types and shadows such as we have in the OLD and NEW Testament Scriptures.

But here you may ask, "What is the necessity of any revelation at all?   Why should mankind have needed a revelation either direct or indirect, either open or symbolic?"   I have answered this question by implication in a dozen places, but I am glad of the opportunity to give it a full and explicit solution.   I shall proceed to do so in my next Letter.

Yours truly,

—————.

# LETTER XVIII.

*Paris, Jan. 12th,* 1857.

My dear W.

Your present intellectual demand may be interrogatively expressed thus: *Why does the Divine creation involve the necessity of a revelation ?* This is only asking in other words, what is the scientific force of the *logos,* or creative WORD: or still again, why has the *Church* hitherto been the leading feature of human history. I do not profess myself to be an adept, but only a learner, in these sublime fields of inquiry, and I am besides subject to a painful suspicion that I do not concisely report what I clearly enough apprehend : yet as the faintest exhibition of truth is powerful to dislodge error, I have no doubt that I shall be able to satisfy your reasonable demands in some sort, if I am only sustained by your cheerful attention and goodwill.

The entire philosophy of the creative *Word* or *logos,* is to be found in the fact that the creative *nisus* is not physical but purely spiritual : in other

words, that creation is never an absolute but
strictly a rational—never a wilful but strictly an
orderly—procedure on the part of God, involving
a due adjustment of ends to means and of both to
effects.  But all this is too succinct for use without elucidation.  Let me rather say then that the
reason why creation seems to involve a previous
literal revelation of the Divine name, or implies
both a letter and a spirit, is that God creates *only
forms or subjects of life,* never life itself.  Properly
speaking, creation is always a redemptive process,
consisting in the bringing life out of death, good
out of evil.  The Divine creation, in other words,
always *pre*-supposes a field of existence, or consists
in God's giving real being to what in itself possesses only a seeming being.  It is a logical contradiction to suppose God giving being to what
does not even exist, to what does not even *appear*
to be, or to what is destitute of self-consciousness :
because this would be tantamount to supposing
that He gives being to nothing, which is denying
creation.  Nothing does not exist.  To suppose
God giving being to nothing therefore, to what
does not exist, to what does not possess even an
*apparent* being, is precisely the same as affirming
a realm of non-existence within the universe of
being, that is, within the scope of the creative
operation.  It is to make nothing convertible with
something, non-existence equivalent to existence :
which is the mere wantonness of absurdity.  Only

things exist, and nothing has no existence. In denying then that God gives being only to what exists, only to conscious or visible existence, we virtually affirm that He gives being to nothing, and so deny creation altogether. The old orthodox theological formula does indeed say, that *God creates all things out of nothing :* but this only means that no mere *thing* is the creative source of any other thing, much less of all other things : or that all sensible existence claims a strictly spiritual and supersensuous being. It by no means professes to be a scientific appreciation and statement of the facts in the case.

No, as Swedenborg shews in his pregnant little treatise on the *Divine Love and Wisdom,* God creates *ex vi termini* only forms or subjects of life in which He may dwell as in Himself. He cannot, as we have seen, create being of course ; because " creating " means *giving being,* just as " making " means *giving form,* and what arrant nonsense it would be to say that God gives being to being ! From the very necessity of the case He creates or gives being only to subjective forms, that is, to organized or conscious existence.* Thus all true creation implies a subsidiary process of formation or making. *And God blessed the seventh*

---

* " Absolute being " as it is called, is unconscious or inorganic being, is being without form ; and unformed being is non-existent being, for existence means the going forth of substance into form, means, that is to say, a process of formation.

*day and hallowed it, because in it He rested from
all the work which He* CREATED TO MAKE.  That
is to say, when the creature is properly made or
formed, and not before, the true divinity of his
source unequivocally avouches itself: then, and
not till then, God's rest or sabbath in him is accomplished, for then first He is able to fill him
with the fulness of His immortal innocence, and
with the exhaustless power and peace which that
innocence conveys.

Now we have already seen that the human form
or selfhood alone is equal to this great destiny,
because it is the only one in which the universal
element serves the individual one.  In the human
form alone the feminine element (meaning by that
term whatsoever is internal, spiritual, and private)
transcends the masculine (meaning by that term
whatsoever is external, natural, and communistic).
Thus the human form is essentially spontaneous
or free.*  Man alone possesses spiritual indivi-

---

* I use these words synonymously, because we know no positive freedom which is not convertible with spontaneity.  Our
spontaneous force argues a complete accord between life and
existence, between what is real or essential in us and what is
merely phenomenal or constitutional, that is to say, between the
private individual selfhood and the common universal nature.
It springs indeed from the perfect marriage-fusion, or unity, of
our spiritual with our natural parts, and is aptly typified by all
the wealth of that angelic bond.  Thus in spontaneous action the
outward or natural individuality freely obeys the inward or spiritual one, finds life and delight in doing so: just as in the angelic
marriage the husband is secondary or passive, and the wife pri-

duality, or as it is commonly called, private character, because he alone is able to postpone his natural appetites to his individual attractions, or bend his common instincts to the service of his private tastes. In a word, the distinctively human form is that in which the instinctual or common life serves only as a basis to the æsthetic or individual one. This is what makes it truly image the creative perfection. God is a universal creator: *i. e.*, He gives being to all things, to whatsoever exists. Here you observe that the individual element *(God)* is primary and controlling, while the universal element *(all things)* is secondary and derivative. In fact, you observe that the Divine subjectivity involves the universe of existence, and hence disclaims any outward object. Accordingly the only fitting form for the Divine influx and inhabitation, must be one of a like universality, or must combine its constituent elements precisely in this manner, always exhibiting its private spiritual or individual element in a controlling atti-

mary or active. (See *Appendix C.*) In instinctual action, on the other hand, the masculine, or outward natural, principle dominates the feminine, there being nothing but a natural individuality known to the animal form. In moral action again the case is reversed, for here we see the female or spiritual element coercing the male or natural one into its subjection. Instinctual action characterizes us in so far as we are still subjects of nature, still animals. Its moving spring is necessity. Moral action characterizes in so far as we are still subjects of society. Its moving spring is interest or duty. Spontaneous action characterizes us in so far as we are delivered from the subjection of nature and society by coming into the subjection of God. Its invariable motive is attraction.

tude towards its public, natural and common one.
Now in man alone, as we have seen in former
Letters, is this requisite imagery fulfilled, for
man alone has a universal subjectivity, or finds
the realm of sensible existence embraced within
the grasp of his proper consciousness. The realm
of infinitude or of the not-me, which is the strictly
spiritual and objective realm whence descends all
our real individuality or character, falls exclusively *within* man, and is never sensibly but only
rationally cognizable. This accordingly makes
the eternal distinction of man, that the entire
sparkling and melodious universe of sense is but
the appanage of his nature, is but the furniture of
his proper life, is but the platform of his true individuality, while the source of that life or individuality is itself for ever hidden in the inscrutable
splendours of God.

But now we know very well that this true self-
consciousness of ours, this distinctively human
form in us, is for a long time imprisoned in its
mere physical conditions, is long immersed in
mere animality. During the immaturity of the
scientific intellect, or his rational nonage, man
regards himself only as a higher product of nature,
only as a superior kind of animal, and never
dreams of associating himself spiritually or inwardly
with Deity, but only naturally or outwardly.*

---

* The orthodox theology during this period represents God as
an outward person, finited in time and space, every way able and
willing to reduce us to an exclusive regimen of kicks and coppers.

His intellectual elevation out of these basenesses
is altogether contingent upon the rise of a scientific society or fellowship among men, upon the
spread of the sentiment of human unity or brotherhood. So long as this sentiment is purely
instinctual, being bounded by the ties of consan-
guinity or neighbourhood, so long of course society
remains without any scientific basis, and does
extremely little for human development, does in
fact almost nothing towards putting man in free
relations with his kind. But society finally out-
grows this natural cuticle. Man gradually learns
to recognize all men as his brethren or equals,
and grows ashamed of loving his father and mother, his neighbour and fellow-countryman, with
a love superior to that which he accords to all

---

The Pantheistic amendment of the orthodox conception still more
hopelessly finites the Deity, by identifying Him with the totality
of time and space, or the entire realm of the finite. It is as if
you sought to aggrandize your friend by resolving him into his
elongated shadow. I confess that if I were driven to choose
between Orthodoxy and Pantheism (instead of saying, as I now
cordially do, " a plague o' both your houses "), I should greatly
prefer for my own worship a being of the utmost orthodox
leanness, to one so intolerably stuffed, plethoric, and wheezy
as this Pantheistic deity must necessarily be. I have the greatest
personal respect for the cultivators of that luxurious creed, but
I cannot conceal my persuasion that the soul invincibly repugns
the bare conception of a God, of whom stinking fish, addled
eggs, and all the other phenomena of corruption, enter necessarily
into the constitution, or even into the authentic though partial
revelation.

other men. He learns at last to love his kindred
and neighbours no longer for their relative or negative worth, but only for their positive and human
worth : no longer for *what is their own* in them,
and therefore separates them from the rest of
mankind, but only for *what is God's* in them,
and therefore unites them with all other men.
In short, instead of any more loving himself in
his friends, he begins to love humanity in them,
esteeming those his truest relatives and neighbours
who most relate him, or bring him nighest, to
universal man.   This is that irresistible sentiment
of human brotherhood, the outgrowth of our scientific culture, which is the vital source of all our
present wide-spread ecclesiastical and political dis-
organization; and when mankind shall have become sufficiently leavened by it, it will compel
society to lift all her members out of the abject
and shameful want in which so many of us still
grovel, by ensuring us all, without distinction, a
comfortable physical subsistence, or a supply of
our absolute physical necessities; so permitting us
for the first time to draw a veritably free and
human breath, and realize our inward alliance
with God.

But now observe: so long as this beneficent
social destiny of man remains unaccomplished, so
long of course our distinctively human force, our
true self-consciousness, remains completely sub-
merged by the natural one; all that is manly,

free, spontaneous in us being held in abeyance to
our basest physical necessities. And *equally of
course*, therefore, *the Divine life in man* (which is
a spontaneous life*) *is meanwhile denied any orderly expression, is without any just scientific ultimation*, being obliged to clothe itself in purely
figurative drapery, or bury its benignant human
meaning under a thick and cumbrous veil of
typical rites and ceremonies. This obligation follows from the very definition of spontaneity, and

---

* The perfect or Divine life in human nature, as we have before
seen, is a spontaneous life, or one which *interiorates* object to
subject. Its subject obeys a wholly inward attraction, renounces
all outward objectivity or inspiration. This life is perfectly symbolized in the historic incidents of the birth of Christ. He was
born of a virgin mother, of a woman who had never known man,
being conceived of the Holy Ghost. This virgin mother, bringing
forth fruit to the Divine Spirit, signifies our natural selfhood released at last from the despotism of the finite, from the long
tyranny of outward want, and quickened exclusively by God, or
from within. The virgin is a beatified Eve. She is Eve emancipated from the coarse Adamic thraldom, and accordingly represents the human selfhood no longer servile to the selfish lusts
which spring from the penury and compression of nature, but joyfully responsive to the inspirations of its inward freedom, and
fruitful therefore of every Divine word and work. The recent
outburst of *Mariolatry* in the Romish communion, stupid enough
when viewed as a rational fact, is yet not without a certain scientific interest in a symbolic point of view. It looks as if our interior Divinity, tired of waiting for its true and perfect expression
in a beautiful life of man, scientifically redeemed from want and
ignorance, or elevated into the universal fellowship of his kind,
sought once more to bring itself to human recognition, by inflat-
ing the old and deceased symbols.

I beg your pointed attention to the observation, for unless you clearly apprehend the truth I am now enforcing, you will infallibly miss in my judgment the whole distinctive scope of the new economy.

The spontaneous life, as I have just said in the preceding note, is one which *interiorates* object to subject. That is to say, it is a life which necessarily brings the object of all my action, the object of all my aspiration, the object of all my worship, *within the conditions of my own nature.* In short, it is a life which exacts the essential humanity of God, which requires that the Deity I aspire to unite myself more and more intimately with, *should be an infinite or perfect man, in all the length and breadth, height and depth,* of that much misunderstood word. Now such being the true life of man, it must always have existed in a shape proportionate to his consciousness of himself. That is, it must have always existed either in a negative or positive form, either as germ or flower, either as egg or chick. But it does not even yet exist in this latter state. We have not yet attained to our true human consciousness. Individuals here and there dimly discern the Divine seed in them, but the mass of mankind seem utterly destitute of spiritual quickening. Priest-ridden and police-ridden, amidst all God's overwhelming bounties they nourish only the furtive courage of mice, and under the kindling sunshine of truth contentedly maintain the darkened intelligence of owls and bats. It follows, then, that our true life must have hitherto existed only in a germinal or rudimentary form, only in the form of an egg, as it were, out of which in the fulness of time should be hatched the consummate vital reality. And this germ of the perfect life—this rudimental embodiment of it—this sheltering and succulent egg, so to speak—has always been furnished by what we call revelation, or simply religion, or still more simply the Church as distinguished from the State. Some purely spiritual revelation of the Divine name in the individual soul, and *failing that*, some merely ritual and symbolic attestation of it, appears to have been as much a preliminary necessity of our perfected consciousness, as the egg is a preliminary necessity of the chicken, which is for a long time unconsciously housed within its frail transparent walls.

I say a " necessity," and this necessity will be obvious to you when you consider the true scope and meaning of our perfected life, when you consider what is inseparably implied in it. The form of the Divine or perfect life in man, is that of spontaneity or freedom, because it is a life which is developed exclusively *from within to without*, and never from without to within. This is the distinctively human form of life, at all times and under all circumstances, whether man knows it or

whether he is ignorant of it, and it invariably
brings forth fruit precisely apposite to such knowledge or such ignorance. But man's first consciousness is natural, and afterwards spiritual:
that is to say, he feels his common or associated
existence before he feels his individual or private
one. Of course therefore both these forms of consciousness, both his natural and spiritual form,
must reflect the true law of his life, which is freedom or spontaneity. His natural selfhood, his
common or associated existence, no less than his
individual or private one, must *in its own manner*
reflect the human form of life, must image the
great controlling law of freedom or spontaneity.
Otherwise his unity of consciousness, his sense of
personal identity, would lapse, inasmuch as there
could be no basis of continuity between his natural and spiritual existence. In short, the true
and Divine life of man, the life of spontaneity,
must shape his natural development as well as
his spiritual one into conformity with itself: that
is to say, must subject the mind of man in nature
to a strictly *historic* evolution, to such an evolution as makes its highest spiritual or individual
culture to be nothing more than the strict efflorescence of natural or universal germs. Such is
the idea of History. It means efflorescence. It
means the continuity of an identical germ through
root and branch, through stalk and leaf, to fruit:
the procession of life from a hidden or invisible

seed to a gorgeous and kingly flower fit to illustrate the sunlight.   In fine, it means the growth
of selfhood.

But you have enough now to think of 'till the
next Letter.

Yours truly,

———.

## · LETTER XIX.

*Paris, Jan. 20th,* 1857.

MY DEAR W.,

You complain of my last Letter as insufficient. It could not very well be otherwise, seeing that I had not bargained to send you a volume of well-digested metaphysics, but only a friendly and suggestive Letter. Let me endeavour now to resume the same theme in a form somewhat more expansive.

You know that ninety-nine persons out of a hundred (and this is speaking with exemplary moderation) envisage creation as a question of time and space—as, at most, a series of sensible facts or incidents, like the American Revolution—and as essentially involving therefore no considerations beyond the ordinary collation and discrimination of evidence. The mass of people believe that creation took place "once upon a time," somewhere in Asia probably, and was complete on the instant by an exertion of physical energy on the part of the Creator. They suppose that some

six thousand years ago, more or less, man was
effectively created, and that his entire subsequent
history consequently has been little better than
a vigorous and unaccountable kicking up of his
heels in his Creator's face. The abject childishness of this conception fails to strike them, only
because the application of reason to sacred subjects has been so effectually discouraged by the
clergy, that our popular intellectual stomach has
grown indurated and ostrich-like,—stowing away
all manner of innutritious corkscrews, jack-knives,
and rusty nails, which may be presented to it by
its lawful purveyors, as if they were so much
reasonable and delectable Christian diet. Indeed,
if you commit yourself to the orthodox conception
of the Divine name, you have no right to denounce
such a diet as unreasonable. A faith full of re-
volting difficulties is a logical necessity of the
orthodox conscience. It prefers such a faith to
one from which all rational contradiction has been
studiously eliminated. For, having no strictly *human* conception of God, having only the *personal*
conception which allows Him to be (at least in all
*practical* regards) a supremely wilful arbitrary and
disorderly being, intent upon forcing all things
into his allegiance and crushing what cannot be
so forced, the orthodox worshipper can of course
conceive no homage half so propitiatory toward
this terrible power, can contrive no flattery half
so subtle, as that which lies in pain and anguish

I

of body and mind voluntarily incurred for its
sake.

Regarded from any such point of view, creation
incontinently tumbles into a rational absurdity or
contradiction, driving us to infidelity and atheism
as to a plain intellectual obligation, as to the only
bed capable of refreshing the weary harassed soul.
For, as Swedenborg declares, so long as we regard
creation as a mere physical event, or as a phenomenon of space and time, we fail to discern it
altogether : and what we altogether fail to discern
by the understanding, we certainly cannot admit
to be true.   The truth is indeed exactly opposite.
Creation is never a mere physical performance on
the part of God, or an event in time and space,
else hounds and hares, cats and rats, spiders and
flies were as authentic creatures of God as man
himself.   On the contrary, it is a purely spiritual
process, falling wholly within the sphere of consciousness, that is within the realm of affection
and thought; or what is the same thing, depending
for its truth upon the evolution of the human form,
which is the sole spiritual form known to the
universe.   It is not possible for God to create, or
give being to, hounds and hares, cats and rats,
spiders and flies, because these things are utterly
devoid of spiritual consciousness.   They are strictly
animal forms, in which the feminine or individual
element is completely controlled by the masculine
or universal one; and God cannot possibly dwell

in, or give being to, forms so remote from His own image, so incapable of free or spontaneous action. To suppose Him inhabiting such forms would be, analogically, to deny His strict objectivity to the universal consciousness, and affirm in lieu thereof His strict subjectivity : would be, in plain English, equivalent to denying that all things were subject to God, by making God subject to all things. He creates only man, who is above all things a spiritual form, a form of spontaneity or freedom exactly proportionate as we have seen to the Divine form, because in him the individual or feminine element is internal and superior, while the universal or masculine one is external and inferior. Only in such a form may God "dwell," to use Swedenborg's phrase, "as in Himself." He truly vivifies only the virgin selfhood, the selfhood which has been released from the bondage of the finite, or from all physical and social compression, and obeys the sole voice of attraction, the inspiration of what we call ideas, meaning thereby infinite or supersensuous good. When my individuality transcends its wonted physical and moral anchorage, when it soars away from the servile earth of necessity and duty into the clear majestic heavens of spontaneity or freedom, it then obeys its essential spirituality, it then becomes feelingly immortal, I then feel the interior and inseparable Divinity of my source, and for the first time taste the rapture of deathless conjunction with infinite goodness,

truth and power.  What does the hare know of
this experience? or the cat, or the spider? Simply
nothing: because they are all alike spiritually
incompetent, being all alike void of spiritual consciousness, all alike incapable of transcending the
natural plane, and allying themselves with infinitude.  I am capable as man of postponing appearances to realities, or of preferring an infinite good
to a finite one.  I am capable of hating father
and mother, brother and sister, wife and child,
lover and friend, home and country, in pursuit of
an interior ideal object, or whenever these base
actualities claim to separate me from that infinite
Divine reality which is the inmost life of my life,
the inextinguishable bliss of all my being.  But
the hound will never know a superior inspiration
to that which his nature devolves upon him, as
it devolved equally upon all his forefathers; nor
the spider ever conceive any bliss comparable with
that of fly-catching, which has descended to it from
a lineage so bloodstained and immemorial, as to
make your ruddiest English pedigrees look pale and
cheap and modern in the comparison.*

* No English nobleman can possibly be as thoroughbred as
the rat which burrows in his own ancestral walls; because, let him
do what he will traditionally to paralyze the human or spiritual
force in him, his bare natural form perpetually prevents his
lapsing into animality, by itself allying him with God, so forbid-
ding him to remain the mere child of his father.  The nobleman
of to-day, whatever be his private vices, is vastly nearer the human type than the nobleman of five centuries ago, simply because

So far then from looking at creation as a Divine improvisation, as at best a mere initiatory incident of history, we are bound to turn the tables and look upon history itself as a mere initiatory incident of creation. If you posit creation as a physical event, as an event of time and space; if you reduce it in short to the dimensions of nature; it is still most incomplete, and all our past history with its lively disputes of Atheist and Deist, of believer and sceptic, is but the flagrant witness of this incompleteness. Who can imagine scepticism existing in the presence of a really Divine creation? In view of a creature visibly vivified by infinite Love, who can conceive of belief as driven to suspend itself upon a laborious balance of probabilities? Our historic experience in fact is nothing

his very nature itself is progressive, while the animal nature is not. For man's natural form being itself spiritual, is incessantly created, vivified, quickened, inhabited by the Divine, and hence is essentially progressive. On the other hand the rat of to-day exhibits not a whit of natural advance upon his antediluvian progenitor, nor ever will, simply because he *is* a rat, and therefore divinely uninhabited or uncreated and consequently unprogressive. Spiritually or interiorly viewed, the whole pretension of an hereditary aristocracy is to animalize the human soul, or dissociate man from his divine original, by making him a creature of *bloods*: than which there can be no profounder blasphemy. This is the secret of those apparently dying throes with which all Christendom is now politically agape and aghast. We are at a crisis in the life of humanity, one of those periods in which man is providentially summoned to shed his old skin, and put on a new one, more pliant to the behests of his inward and essential freedom.

but our gradual approximation to human consciousness, and to the consequent consciousness of ourselves as Divinely created. It marks nothing but the endless interval which separates the highest animal form from the lowest human one. We have indeed no business to look upon human history as an accident, as a something *supervening* upon our creation, as a direction impressed upon us by some power extraneous to our nature. On the contrary it is a most strict incident of our creation, being nothing more nor less than the ceaseless effort of our essential Divinity to give itself adequate formal utterance or embodiment. God is essential man, and human history is but the gradual adaptation of this superb spiritual truth to the natural imagination of the race. All its sacredest incidents accordingly, far from denoting any outside interference with our nature, are the strict outgrowth and efflorescense of august interior powers. Thus what we call a Divine revelation, what we call religion, or the Church, is never an arbitrary external imposition upon the human mind, but on the contrary is always a normal though fruitless effort of our interior Divinity worthily to assert itself in the plane of the senses, or to attain to scientific recognition. It is in every case the Divine or spontaneous life of man seeking to secure itself a representative or figurative projection, so long as it is denied a living or conscious one. In short, history, strictly speaking, is our process of FORMATION. It is the untiring effort which the creative Love makes to bring us up to the human form, to develope in us spontaneous life, to endow us with a selfhood adequate to image its own perfection, and therefore adequate to its own indwelling: and all its successive stages mark only so many successful crises of that effort.

Let us then boldly reverse our point of view. Let us cease to regard creation as an historical incident, as an event in time and space, by learning to regard history itself, or all the events of time and space, as mere incidents of creation. History, I repeat, means nothing else than the evolution of that distinctive human form which belongs to us as veritable creatures of God, as beings vivified by a really infinite breath, by a really perfect power. It is the gradual vindication of a Divine NATURAL humanity. It is in a word our needful natural formation in the Divine image. The fundamental import of Christianity, the fundamental import of all authentic Divine revelation, is, that we need to undergo a natural formation in the Divine image in order to our spiritual creation; that our spiritual or individual creation by God really exacts for its own permanent basis our natural regeneration. The religious idea, separated from the caricatures of superstition, implies, that it is incumbent upon the Divine bounty to give us natural selfhood quite as much as spiritual selfhood; that unless we first bear a common or associated

likeness to the Divine, we shall be destitute of a
private or individual likeness. The ground of this
exaction lies no doubt in the great law so often
cited already, that God creates only subjective or
spiritual existence : but you will not be prepared
to do justice to this law, or accurately to comprehend its bearings, so long as you cherish vague
and obscure conceptions of what is meant by
creating. Let us manfully free ourselves of the
stifling traditional nonsense on this subject, and
then we shall perfectly understand why we require
to be naturally as well as spiritually fashioned in
the Divine image, or what is the same thing, why
a Divinely-given natural form is an indispensable
preliminary basis to our Divinely-given spiritual
being. And, understanding this, we shall have an
infallible clue to the religious history of the race,
which is the veritable history of the human mind,
and be able clearly to conceive why that history
intimately involves the doctrine of a Divine revelation or incarnation.

Let me beg of you then distinctly to remember
that I use the word *create* with strict scientific
accuracy, as always meaning *giving being*. To
create a thing means to give it inward or substantial being; he who creates a thing *himself constitutes the substance* of that thing: so that the
relation between Creator and creature is invariably
the relation of object and subject, of internal and
external. Creating or giving being is an exactly

inverse process to that of making or giving form.
When I say that God creates me, I suppose myself
already formed or existing; I take my existence
for granted, or as inseparably implied in my proposition. Existence is an absolute and indisputable
fact, and unless we had this preliminary basis of
sensible experience, we should be utterly void of
supersensuous experience of every sort, whether
belief, or hope, or aspiration. Accordingly in
alleging my creation by God I do not refer to any
mere fact of existence, to any sensible operation of
God, but wholly to a spiritual and invisible operation; one which utterly transcends the realm of
time and space, because it falls altogether within
that of affection and thought. In other words,
in alleging my creation, I do not project myself
back in imagination to some period more or less
remote, when an exertion of voluntary energy on
God's part resulted in my physical genesis or
formation—resulted in giving me existence. Far
from it. I take my physical formation or existence
*pro confesso*, as an indispensable platform of the
creation which I allege. For I say that God
creates *me*, and obviously by *me* I mean my human
form, my phenomenal existence, my conscious personality. It would be absurd of course to allege
any abstract creative energy on God's part, to say
for example that He creates what has no existence,
or what is unconscious and invisible: because, as
we have already seen, that would be only saying

in a round about way that He creates nothing, or
that He is no creator. We can never conceive of
creation except as proceeding on the basis of some
existing selfhood, as involving some subsidiary
sphere of formation, as predicable in short of
certain conscious or visible existences. By saying
that they are created existences, we do not mean
to allege any physical fact whatever concerning
them, but on the contrary a purely metaphysical
fact, which is, that their being is not identical
with their visible form or existence, or, what is the
same thing, that they as subjects involve a far
profounder objectivity than that of nature. And
by saying that God creates them, we mean that
He who is infinite Love and wisdom constitutes
their spiritual and invisible being : that He stands
to them in the eternal relation of inward genetic
source or object, and they to Him in the eternal
relation of outward derivative stream or subject.

You may doubtless ejaculate a ready *Amen* to all
this, by way of inducing me to resume my initial
proposition, which is : that God creates only spiritual forms, gives being only to subjective existence :
but I feel so cordially disposed to disabuse your
excellent understanding of certain sensuous falla-
cies and prejudices engendered by the Old Theology, that I cannot forbear to solicit your indul-
gent attention a few moments longer. I want you
perfectly to comprehend both what is included in,
and what is excluded from, the rational or scientific conception of creation.

Let me distinctly say then, that the technical
infidel is completely justified in denying creation,
so long as you represent it as implying an outward
exertion of Divine power, as meaning a physical
operation of God. The letter of revelation no
doubt represents creation in this guise, that is, as
a simple projection in time and space, as a strictly
· *impromptu* proceeding on God's part, involving
nothing more than a new determination of His
will, and the consequent utterance of an authoritative *fiat*. But all this is a purely symbolic or
pictorial statement of the truth, without the
slightest value as history. If indeed you view
it as literal history, it becomes at once downright
puerility and nonsense, since it represents God as
creating mere natural existence, or as being simply
what is termed "the author of nature," which is
totally to degrade His name, and render it the
inevitable butt of the flimsiest sentimental devotion, the tattered target of the mildest Unitarian
archery. Natural existence is absolute existence,
being that in which substance and form are identical. Nature means the identity of substance and
form, of being and seeming. The stone for example, the tree, the horse, *is* exactly what it *seems*
to your eye. Its being is a pure seeming, is wholly
phenomenal, as the philosophers say. There is no
spiritual stone, nor horse, nor tree, lying back of
and animating the apparent one. The sensible
form before you perfectly embodies its own being

or substance, so that every stone, tree, and horse
of the specific family in question will repeat the
same monotonous story over again till time and
space shall be no more.  You can't imagine a stone
or tree, or horse, out of relation to time and space,
that is as having any purely subjective or spiritual
existence by virtue of its inward commerce with
infinite goodness and truth.  You can only conceive of them as natural existences, thus as essentially finite and perishable.   Observe then that
natural existence is purely phenomenal existence,
being destitute of internal or individual being and
hence out of all immediate relation to God.  Yet
this is the prevalent conception of creation, the
only conception tolerated by the carnal or superstitious mind.   And what is very melancholy, the
clergy as a body do their best to confirm and
aggravate our natural hallucinations on this and
every subject.  They are wont, as a general thing,
to attribute to God the dreariest and most tedious
existence imaginable, by diffusing His infinitude
over the wilderness of space, and trickling His
eternity through the endless succession of minutes
which make up time; and then they represent
Him as suddenly resolving to variegate this barren
infinitude—to diversify this monotonous eternity
—by summoning into life certain absolute or physical forms, which shall henceforth be and exist by
virtue of that momentary *fiat*.  In short the ecclesiastical intellect all the world over has the invete-

rate habit of confounding being with form, creating
with making, reality with semblance. It supposes
that every thing really *is* which *appears* to be: or
that things have *being* by virtue of their *form.*
If for instance you should consult the Pope of
Rome or the Archbishop of Canterbury, they would
never betray the slightest distrust of their official
existence being a Divine reality. They have not
the least suspicion that the higher powers are
blessedly ignorant of all the conventional dignities
of the earth; they have never imagined that all
those distinctions, official and personal, which
make up so often our best knowledge, and give
many an empty head among us the reputation of
wisdom, are sheer vacancy to the celestial mind,
raying out darkness, not light; and if you should
hint your own suspicion of the truth, they would
cordially unite in proclaiming you an infidel, and
bid you begone as a tiresome revolutionary bore.*

* I feel no *positive* admiration for the revolutionary forces
which are now enthroned in France, and only waiting to be
effectually enthroned over the rest of the European continent;
because I see that they are mere Providential tools employed to
work out far diviner ends than they themselves dream of. But
when one reflects upon the crowned imbecilities which actually
rule over men, sacerdotally and secularly : when one considers the
fearful distance which separates the conventionally upper classes
from the lower; their utter aloofness from the common loves, the
common wants, the common hopes of man; their luxurious self-
indulgence; their unrighteous social privileges, and the inevitable
pride and arrogance engendered by such privileges; their stolid
opposition to popular elevation ; their hardened indifference to the

But there is no need of troubling Pope or Archbishop with these inquiries, especially as they have already trouble enough on their hands, I dare say. Suppose the question put to you, John Doe, and to me, Richard Roe: "if the visible selfhood we are each of us born to, be indeed the vital reality which it seems to us to be :" we should unhesitatingly answer, Yes. You have an undisturbed conviction that you are *personally* known to God, that your luxuriant locks, your dark eyes, your tint embrowned by sun and air, are perfectly familiar to the Divine eye. And I for my part have never questioned that the Divine mind was as cognizant of my visible limitations (short stature, obese figure, fair complexion, flaxen wig, and so forth) as I myself am. Yet this is a sheer mistake. Swedenborg, who had a great eye for realities as discriminated from mere appearances,

voice of God's great minister, science; their flippant contempt of every force but brute force, and their inveterate estimate of humanity as an essentially brute existence, never to be regulated from within, or Divinely, but only from without, or diabolically : then Louis Napoleon, Mazzini, and all the rest, become irresistibly precious and sweet to my heart, even as terriers and weasels are precious to the agriculturist long vexed by predatory and fugacious vermin, even as the advent of death's angel is sweet to the soul long imprisoned in a diseased and suffering body. In fact one respects the Revolution very much as one respects Death. It is not in itself a Divine presence any more than the rotten and odious *régime* which it has displaced; but it constitutes the only door which our double-dyed stupidity and unbelief will ever leave open to the entrance of the Divine kingdom on earth.

could never find a vestige "of the old familiar
faces" beyond the grave. The phenomenal selfhood was fatally transfixed and dissipated by the
first contact of trans-sepulchral light. He knew
many. persons of a very conspicuous conventional
make, heroes and saints, statesmen and clergymen,
abounding in learning and piety; but when he saw
them illumined by celestial light, he frequently
found them full of rapacity cruelty and excess of
all sorts, and degraded to the most menial positions.
And so, on the other hand, he not unfrequently
found persons, who on earth and to their own
consciousness were destitute of every claim to sanctity, who lived in affluence and luxury, who fre-
quented theatres, who loved jocose conversation,
who had in short no properly *ascetic* fibre in their
composition—mere unbaptized Turks and Pagans
very often in fact—enjoying an intimate commerce
with the angels, and heartily allied with all Divine
perfection.

All this (and very much more) is true, I say,
simply because the phenomenal is never the real,
because what *appears* never *is*. The sensible world
is purely formal, not essential: it is, and ever will
be, the realm of shadow, not of substance; of
seeming, not of being. It is not the theatre of
the Divine creation, but of the Divine formation
exclusively, being, to use Swedenborg's phrase, a
sphere of effects not of ends. In short, Nature
.is a purely *experimental* world, and experience is

a first-rate mother, but a most incompetent father.
Experience incarnates our wisdom, or gives it
outward body : it does not vitalize it or give it
inward and rational soul as well.  In all procrea-
tive action the father is generative, the mother
simply prolific or productive : the former gives life
or soul, the latter existence or body : the one is
creative, the other formative.  And this diversity
of function is but an image of the universal spiritual truth, that experience (or our natural memory) serves only as a ground or matrix, only as
a warm mother-earth, in which to inseminate certain formal traditions, which are the mere husks
of truth, inherited from the past, while God alone
(or Infinite Love *within* the soul) constitutes the
stainless overarching heavens by whose genial
beams these rude and lifeless husks become quickened into every form of living wisdom.  We know
that every seed must die in order to bring forth
fruit.  All food must be dissolved before it can be
assimilated, before it can make flesh.  Now these
natural facts are but the shadows of spiritual things.
All the literal dogmas we receive into the memory, which is the mental stomach, are of no
more promise in a spiritual point of view, than
so many stones taken into the natural stomach
would be in a hygienic point of view.  They give
us hope of spiritual increase only in so far as they
undergo intellectual levigation or maceration, only
in so far as they become converted into that rich

rational chyme and chyle whose white depths
nourish and embosom the immortal pillars of the
soul.* Understand then that Nature is the realm,
not of wisdom, but of that experience which is the
indispensable soil of wisdom. It is the sphere not
of soul, but of that needful preliminary bodily
organization without which the soul itself would
never come to consciousness. God cannot directly
create natural things therefore, because these
things, being fixed or absolute, forbid that interior
expansion, that perfect individual freedom, which
is the inseparable heritage of His creatures, and
which alone conjoins them with Him. The horse,

---

* This is what makes mere *professional* religionists so tiresome.
For having not merely the ordinary human but also a distinctly
private or personal end in the maintenance of our traditional
creeds, they sedulously guard them from all intellectual fecunda-
tion, from all rational trituration and fermentation, and hence
perpetually suggest to the imagination the painful similitude of
people in a colic. They present the same contrast to our ordinary
unconscious and placid acquaintance, that the shop of a seedsman
and florist presents to a blooming and beautiful garden. In the
professional religionist, the memory is sure to grow plethoric at
the expense of the reason, just as we often see a man cultivating a
portentous abdomen to the serious neglect and discredit of his
brain : and intercourse is never at its just human pitch, until it is
above all things rational. When our intercourse is one of cant,
being vitalized only by the memory ; when, in other words, my
friend and I meet only to parade and compare our mutual wealth
in current orthodox coin, the image we project upon the spiritual
sense is that of two foolish persons diligently rubbing their sto-
machs together, or belching in each other's face, in order to
inflame a reciprocal good understanding.

the lily, and the diamond, are beautiful natural
existences, but how impossible to fancy them in
any relation to God, simply because though they
have each a marked natural individuality, they are
yet all alike destitute of spiritual or real individuality: in other words, because, though they are
all subjects of a beautiful existence, they are none
of them subjects of life.

This explanation ended, I am now ready to
resume my initial proposition, which was, that
God creates only subjective or spiritual forms.
This follows, almost obviously, from the definition
of creating; for as creating always means, when
properly used, the *giving being* to things, so consequently God can only create or give being to
things which are in themselves destitute of being,
having at best but a subjective semblance or appearance thereof. He cannot possibly give being
to what already has being, since this would be
contradictory, but only to what appears, only to
what seems to be, that is, to subjective or spiritual
existences. I repeat, then, that by the strict necessity of the case, God creates only subjective
spiritual forms, in which He resides as in Himself,
so and not otherwise communicating life.

Now the condition of subjective or spiritual existence is, that it be vitalized from *within*, or what
is the same thing, that the object it obeys, the
ideal it serves, reside strictly within the limits of
its own nature. Natural existence is the opposite

of this. What the philosophers term "objective" existence, meaning by that word whatsoever sensibly exists, as mineral, vegetable, and animal, is always vitalized from without, that is to say, its objective element is strictly exterior to its subjective one. The mineral exists for the vegetable, the vegetable for the animal, and the animal for man. In short, natural existence is servile existence, finding its proper object or ideal out of the bounds of its own nature. Of course this peculiarity puts the merely natural form of life out of all immediate contiguity to the Divine, by leaving it destitute of internality, of private or spiritual individuality. The horse, for example, who obeys an ideal essentially aloof from his own nature, whose deity in a word is man, is by that fact denuded of spiritual consciousness, of what we call selfhood or character, and hence remains essentially unprogressive or incommensurate with God. He has abundance of physical life, of selfhood or character derived from his natural progenitors, but he has no Divinely-vivified individuality athirst for the fountains of a better life. No sweet radiant Eve grows up in the unconscious depths of *his* bosom, becoming evermore bone of his bone and flesh of his flesh, and leading him to eat of the tree of knowledge of good and evil, that through the disease and death thus revealed he may rise to the experience of immortal peace and joy. He knows, no doubt, the natural love of the sex, or recognizes the partner his nature provides him : but he has no glimpse of the ravishing amplitude of bliss which is spiritually locked up in the conjugal symbol, and which makes the *wife* as contradistinguished from the woman, an exquisite shadow of all that is most intimate, ennobling, and enduring in the ineffable commerce of the Divine and human natures. This experience, I repeat, is denied the animal, because the animal form is vitalized from without, because its objective element is strictly exterior to its subjective element, or in other words, because the ideal it promotes, the object it serves, the deity it obeys, is human and not animal, that is to say, does not fall within the grasp of its own nature.

But the exact reverse obtains with regard to man. The human form is vitalized from within exclusively. The objective element in all human activity will be seen on a fair analysis to lie strictly *within* the subjective one. The ideal which I propose to myself as man, the object I seek to promote in every form of action, in short, the Deity I worship, is always of an intensely human quality, invariably puts on the lineaments of my own nature, and hence my life of necessity becomes evermore beautiful and free, abhorring nothing so much as servility. In a word, man's existence is purely subjective or spiritual, compelling even the infinite Divine perfection into his own natural dimensions before it can win his honest and hearty acknowledgment. What is the inmost meaning and confession of all evil but this ? To the inner or instructed sense evil is only the running away of the fish with the line which binds him to his captor, and is but a surer argument of the skill which is bound eventually to bring him to land. Lying, fraud, adultery, murder, covetousness, are only so many temporary diffractions of the pure and stedfast Divine ray operated by our intellectual opacity and indocility; are only so many incessant and stupid crucifixions, wrought by our infatuated carnality and self-conceit upon that Divine and long-suffering Love which underlies and animates our nature. The horse is destitute of morality because, being a purely outward or natural existence, he must for ever remain incapacitated for that spiritual or subjective freedom of which morality is but the shadow. Morality implies a relation of independence, in so far forth as it is predicable, on the part of the subject towards his nature. But the horse is the abject slave of his nature. Every existence indeed below the human exhibits the complete identity of being and seeming, of substance and form, of soul and body. You, on the contrary, as man heartily repugn such identity. You feel so sure of nothing as that your being will always transcend your richest experience of it, or what is the same thing, that your amplest actual must ever fall hopelessly short of your feeblest possible. The real horse is always

the visible horse, and no lily has being but that
which actually blows in the garden, and fills the
worshipping air with its dazzling sheen.  But the
opulence of man is such, the opulence of God's
true creature, that what is visible of him always
confesses itself nothing, however glorious, while
what is relatively invisible claims to be the only
reality.  Thus the visible man is never the real
one.  The man that veritably *is* never shews himself except by proxy.  The true friend must ever
despair of disclosing the passionate depths of his
friendship, and the genuine lover strives always in
vain to interpret himself worthily to his mistress'
sense.  Though he heap Pelion upon Ossa in the
fond effort to storm the flaming heavens of his
love, and compress them into appreciable measures,
they for ever mock his aching embrace, for ever
falling back into the impalpable abysses of the
infinite.  Such, I say, is the normal state of man.
This is his state, when, being emancipated from
physical and social thraldom, he stands erect in
true human proportions.  He is then a purely
spiritual or subjective form, made conscious of
himself no doubt by the background or basis of
his physical and social organization, but utterly
incapable of identifying himself with that organization.  He instinctively feels himself to be superior to his circumstances, to be dearer to the heart
of God than all that calls itself nature and society
put together, and in the robust confidence of that

intimacy seeks evermore to bring both nature and
society into his own unlimited subjection. And
manifestly all this is true of our human instinct
and experience, only because the human form
alone is divinely vivified, only because God does
literally create us or give us inward being, while
He does not do so to cabbages and horses. He
gives them *outward* being, which is natural existence, and which leaves them destitute of all private
individuality, of all spiritual lift above the dead
level of sense. But He gives us *inward* being,
which is spiritual existence, and which fills us
with a private individuality so pronounced and
expansive as eventually to precipitate Nature,
much as we drop our garments from about us at
night, or rather to transmute her from an all-
enveloping and absorbing egg into the very texture
and substance of the new consciousness, into the
very pith and marrow of the new and diviner
manhood.

Of course then the Divine creation rightly viewed, stamps Nature with a deeper significance than
she herself is at all aware of. While to her own
consciousness she seems absolute and final, she is
nevertheless but the seminary or seed-place of the
soul, the mere husk and tally, so to speak, of those
august interior forces which are for ever shaping
the spiritual universe (or the mind of man) into
harmony with all Divine perfection. Nature is in
short but the perishable body of the imperishable

mind of the race, and we fail to see her in this intrinsically subordinate plight, only because we habitually estimate her by the light which she herself supplies, or what is the same thing, because our reason, in place of being *served* by sense, is actually *controlled* by it. Revelation itself is bound of course to conform its utterances to this natural necessity; is bound to respect the limits of the sensuous understanding in man, under penalty of forfeiting its true character and becoming degraded into mere information. That is to say, the Divine and eternal truth can never reveal itself to sense except in a *symbolic* manner, because if it should attempt to assert itself as a fixed or absolute quantity, the human mind would have no chance to grow, being thus authoritatively robbed of its freedom. In other words, the letter of a Divine revelation avouches its authenticity only in so far as it embodies spiritual or universal truth. The general vague impression on this subject no doubt is very different. It is popularly conceived that revelation is not a symbolic unveiling of truth, addressed only to the spiritual understanding of man, but a literal unveiling of it, addressed to his senses. It is sensuously supposed to be a direct and unaccommodated communication on the part of the creator to the creature, leaving the latter no option but to obey. Thus all the gospel facts, so far from being viewed as the normal natural outgrowth and expression of certain Divine operations

within the universal soul of man, are supposed to
have a purely absolute genesis which discharges
them of all strictly human or scientific validity.
But this is the mere dotage and delirium of sense.
The eternal splendour of the Christian facts lies on
the contrary just here, that what seems personal
and limitary about them is precisely what adapts
them to mask universal truth, or to symbolize the
relations of all mankind to God. They have in
truth nothing arbitrary about them, but are one
with the highest reason, being the outgrowth not
of private causes but of universal ones, of causes
which are as wide as the universe of being. I hold
(perhaps more strenuously than you can at present
imagine) that Christ was conceived of the holy
Ghost, that he was born of a virgin, that he lived
a life of helpless humiliation and infamy in the
eyes of the most reputable persons of his age and
nation, while at the same time he became inwardly
united with the Divine spirit to such a degree as at
length to grow exanimate on his finite or maternal
side, and find his literal flesh and blood becoming
vivified by the infinite Love. But then I cannot
conceive of these things being literally true save
on one condition, which is, that nature be not the
absolute and independent existence she seems;
that she be in fact *the mere shadow or image of
profounder realities*, projected upon the field of the
sensuous understanding. For if nature be a direct
creation of God, if she be an existence fixed by

K

the actual creative *fiat,* then the pretensions of the
Christian revelation are to the last degree absurd:
because the Divine creation once actually posited,
must ever after prove incapable of amendment, or
find itself beyond the need of any officious tinker-
ing. This needs no argument. But if nature be
nothing more than the common or ultimate bond
and covering of the spiritual world, which is the
universal mind of man, just as the skin is the
common or ultimate bond and covering of all the
diversified kingdoms of the body : why then of
course we may regard all natural phenomena only
as so many graduated effects from interior spiritual
causes, precisely as we regard a blush upon the
skin, or a sudden pallor, as an evidence of height-
ened or depressed vital action. And so doubtless
day and night, the succession of the seasons, birth
and death, growth and decay, the subordination
of mineral to vegetable, of vegetable to animal,
and of all to man, *are* so many natural types, are
so many ultimate symbols, of a vast and beneficent spiritual order which is inwardly shaping
the universal soul of man, and which will eventually bring about the perfect reciprocal fusion or
unity of each with all and all with each. But
how to divine this recondite knowledge ! Nature
has as little consciousness of man, as the waters
have of the sun and stars which irradiate their
darkened and tumultuous bosom. Nature herself
therefore is incapable of blabbing the secret with

which she is fraught, or of proving a revelation of
Divine mysteries to the soul, because she is utterly
unconscious and incredulous of Divinity. She has
no more comprehension of the being she images,
than the looking-glass has of the human substance
whose various phenomenality it reflects. She is a
pure surface whose depth or soul is man. No
doubt she will faithfully lend herself to the reflection and illustration of his intimate worth, in so
far as his own intelligence learns to demand that
service of her. But she has no independent power
of origination or suggestion. She feels no fore-
warning of the lustrous use she fulfils, until his
advancing self-knowledge imposes it on her. She
has no clearly articulate speech which she does not
catch up from his commanding accents. In short
she knows herself truly only as the echo of his
majestic personality, and shrinks from nothing so
much as the pretension to lisp even a syllable of
original Divine revelation. Revelation descends
exclusively from the human consciousness, or from
the soul of man to his senses, because man alone
being the true creature of God is alone competent
to reveal Him. In short the true theatre of revelation is not our mere natural or animal consciousness, but our historic or veritably human consciousness. It demands for its proper platform
not merely that humble field of relations which
man is under to his own body, and which constitutes what we call his *existence*, being all compre-

hended in the fixed quantity denominated Nature : but also and above all that superb field of relations which he is under to his own soul, or to God, and which constitutes what we properly term his *life*, being all comprehended in that great unfixed quantity which we denominate History.

Only one more letter, and I shall have done.

Yours truly,

——.

# LETTER XX.

MY DEAR W.,

I DO not know how it strikes your intelligence, but it appears to me that I have to some extent indicated in my last Letter the true ground of the difficulty men have in rationally conceiving of the Divine Incarnation. Let us recall for a few moments what has gone before, in order that we may the more clearly take the final step.

We have seen that Christianity abolishes the Pagan conception of Deity, which represents God as an essentially arbitrary, insane, or inhuman, force,—capable at will of any amount of deviltry and destruction,—by revealing Him henceforth as a glorified natural man, as a rightful and permanent denizen of human nature. In other words, the service which Christ rendered humanity—a service to which there has been, and, in the nature of things, can be, nothing similar or second—consists in this: that He furnished by His life of unparalleled self-denial a perfect natural embodi-

ment to the Divine Love: that He shut up the
infinite and hitherto inconceivable Divine within
the dimensions of the humblest of human bosoms;
constraining it thenceforth to know no other activity but that which is supplied by the intelligible
forms of human nature, that is to say, compelling
it to run henceforth eternally in the familiar mould
of our natural passions and appetites.   Let there
be no obscurity upon my meaning.   I say that
what Jesus Christ did to entitle Him to our eternal and spontaneous homage, was that He, by His
unflinching denial, even unto death, of the popular
religion of His nation (a religion which, as to its
*fond*, was fed by every infernal influence, and as to
its form, by every celestial one), He, for the first
time brought the infinite creative love into perfect
harmony with the individual bosom of man—into
complete and unobstructed *rapport* with the finite
human form—so that Deity might once for all
experimentally know how it felt to be husband and
father, lover and friend, ruler and teacher, patriot
and citizen, under that base natural inspiration
merely; and so knowing, for ever vivify and re-
deem those finite ties, by the communication of
His own infinite substance.   I, for example, am a
husband and father, am a lover and friend, am a
patriot and citizen, and in all these characters exhibit a much less arbitrary aspect than I should
have done had I lived in the centuries which pre-
ceded Christ.   Why?   Simply because in those

centuries, as Swedenborg shews, the Divine access
to man in nature took place by angelic mediation
exclusively; and this mediation, being perpetually
obstructed and enfeebled by the antagonism of the
hells, the consequence was, that every natural tie
of man was practically fast becoming a channel of
unmixed selfishness and tyranny. We have already
seen that the angelic form is incompetent by itself
to vindicate the infinitude of the creative power,
because it owns no good more decisive than that
which flows from the incessant elimination of evil.
The angel is an imperfect creature of God, is an
incomplete style of man, because he involves a
diabolic antipodes. In other words, *the heavens
are impure in God's sight, and He charges His
angels with folly*, because they are not spontaneously good, but only voluntarily so: that is to say,
because they are good only by the denial of their
nature, never by its concurrence. Accordingly the
angel must always have proved a most inadequate
point of contact between the infinite and finite.
The Divine Love must have always felt itself hopelessly straitened in its approximation to the human
bosom, by the exigencies of a mediation which
never contemplated the reconciliation or co-ordination of self-love with brotherly love, but only its
forcible extrusion and suppression.

The whole problem of creation may be sum-
marily formulated thus: the natural man (or man
in a state of nature simply, without historic ex-

perience) is a form of supreme self-love, and thus presents an exactly opposite aspect to the Divine Love which is incapable of selfish regards: of course then creation must remain an eternal impossibility unless some middle term can be projected capable of reconciling or fusing these inveterate opposites. Now, I say that the angel could not pretend to furnish this requisite middle term, because his entire vitality proceeds not upon the reconciliation of self-love with higher loves, but upon its forcible expulsion, and even, if that were possible, its extinction. But in the bosom of Jesus, exposed through the letter of His national hope to the boundless influx of every selfish lust, and yet persistently subjugating such lust to the inspirations of universal love, the requisite basis of union was at last found, and infinite Wisdom compassed at length a direct and adequate access to the most finite of intelligences. In Christ unfalteringly renouncing His own sacred writings, in so far as they were literal, personal, and Jewish, and accepting them only in their spiritual, universal, or humanitary scope: in His cheerfully submitting to life-long obloquy for this unprecedented manliness; to the scorn envenomed by disappointment of all that was most decent, devout, and respectable in His nation; to the daily derision of that large class in every community, who, not being devout themselves, yet hope to commend their sneaking souls to heaven's favour by blindly doing the dirty

work of the devout, and hastening brutally to
finish what these are sometimes fearful even to
begin; to the contempt of His own brethren and
neighbours; to the constant misconception and
unbelief of His own avowed, and forward, and
foolish disciples; finally, to death itself—a death
from which no element of ferocious cruelty was
absent, which, on the contrary, all hell found a
truly religious joy in promoting: in this sublime
and steadfast soul, I say, the marriage of the
Divine and Human was at last perfectly consummated, so that thenceforth the infinite and eternal
expansion of our nature became, not merely possible, but most strictly inevitable. Accordingly,
ever since that period, husband and father, lover
and friend, patriot and citizen, priest and king,
have been gradually assuming more human dimensions, have been gradually putting on glorified
lineaments; or what is the same thing, the universal heart of man has been learning to despise and
disown all *absolute* sanctities: not merely our
threadbare human sanctities, sacerdotal and regal,
conjugal and paternal, but also every the most
renowned Divine sanctity itself, whose bosom is
not the abode of the widest, tenderest, most patient and unswerving human love.

Now what I shewed in my last Letter was, that
we deny, or misapprehend this Christian revelation,
only because we have the folly to regard space and
time as substantial things, as veritably Divine

ideas, and to look upon nature consequently rather
as the primary than as the intensely ultimate and
subordinate field of the Divine operation. Nature
is in truth but the basement or culinary story of
the Divine edifice; and when we make her primary, or allow her to dominate the house, we of
course degrade the drawing and bed-room floors,
filling them with sounds and odours fatal to every
cultivated sense. Theology and philosophy have
done little hitherto but fill the world with this
odious din and stench of cookery. Obstinately
regarding nature as the final rather than the mediate sphere of the Divine operation, as the real
or substantial world instead of the purely formal
and phenomenal one, they incessantly drown our
rational intelligence in the mire of sense, whence
we have now actually no more lively theologic
tendency extant than Unitarianism, nor any more
lively philosophic one than Pantheism; from both
of which the scientific intellect, heedful of its own
sanity, is bound heartily to recoil, even if the
alternative should be downright scepticism and
atheism.* The new theology and philosophy re-

* Confiding in the fallacious dogmatism of sense (that *old serpent* whose speech is far too subtle and insinuating to be suspected
prior to experience), our theologians and philosophers regard being
and seeming, truth and fact, reason and experience, as identical, and
hence vainly rummage the phenomenal world for an original glimpse
of those lustrous Divine footsteps which fall wholly within the soul
of man, and of which nature herself is at best but the distant reverberation. Nature is but the echo of the soul, and images

verse the spell. They teach us that creation is primarily spiritual and only derivatively natural, thus that the science of nature is rightly comprehended in the higher science of man. "Yes," they say, "cookery is a strict necessity of things, and claims its proper acknowledgment: but it should never be exalted into an end of life. Its sole end is to nourish and prepare the body for the uses of the soul. So also what we call spiritual regeneration is an actual necessity of things, but it is a necessity which belongs wholly to the natural plane of experience. The soul, coerced by the appearances of things, demands it: instructed by realities, disavows it." As long as I am instructed in spiritual things only by sense or appearances, I deem myself an absolute person in God's sight, and look upon all His dealings towards me

nothing therefore of the Divine creation and providence which is not primarily impressed by the soul. Your delicious English landscape, for example, palpitating with its rich subserviency to every human need, reflects a far more evangelical lesson in these respects than the hideous jungles of Asia, or our own unsubdued forests and indolent savannas; because the humanized English *man* has first taught it so to do. Abstract this comfortable Christian English soul, who believes in nothing more soundly than a deity favourable to good cheer, prolific of everlasting cakes and ale, and your peaceful English landscape would have been by this time as ruthless and unchristian as that of Switzerland, which for the most part suggests no thoughts of Divinity but as of some huge, frowning, thunderous, overshadowing, overbearing power, eternally allied with pride and self-will, and essentially untouched by all those blissful human sympathies and charities whose inseparable root is humility.

as having a most special intention, which is an absolute conversion of me from evil to good. But the reality of the case is, that God never acts upon us individually, save by acting at the same time universally, and consequently that what I regard as a change of nature in me, is in reality a separation of spiritual spheres taking place in the universe of the human mind, by which its external principle (self-love, or *hell)* becomes precipitated, and its internal principle (which is brotherly-love, or *heaven)* elevated, that so the mind of man in nature may be at length effectually harmonized with all Divine perfection.* I feel in myself, for example, a great horror on account of some sin, real or imaginary, which I have committed; I humble myself before God by whatsoever penitential methods my traditional conscience prescribes, having no shadow of suspicion all the while that God is not literally feeling very angry with me, and even extremely dubious whether or not He will pardon me. Such are the crude and abject *data* of my natural experience. But hereupon come the theologian and philosopher, not to give me intellectual elevation out of this superstitious lore, but actually to confirm all its teaching, telling me that my experience is an exact

* Of course it is only when self-love claims the primacy of neighbourly love or charity, that it is contrary to Divine order. When it spontaneously defers to the latter, as it does in the scientific sentiment of human society or fellowship, nothing can be half so orderly and beneficent, and we cannot have too much of it.

measure of the real and eternal intercourse between
God and the soul. They affirm that He is in
truth very much offended with me, just as my still
grovelling intelligence proclaims Him to be; that
I have in fact committed a grievous sin against
Him, and that I only follow the obvious dictates
of prudence in aiming to propitiate Him by every
customary usage of self-abasement. Such is the
help they give my reason, utterly immersing it in
sense. It is as if my cook, in a moment of revolutionary frenzy, should transport his *batterie de
cuisine* into my drawing-room, and insist upon
henceforth preparing my dinner under my proper
nose. For it is really most untrue that God has
ever felt, or ever can feel, an emotion of personal
approbation or personal disapprobation towards
any human being. All this is the mere abject
gossip of the kitchen, the mere idle *bavardise* of
cooks and scullions theorizing in their dim sub-
terranean way upon the great solar mystery of life.
It is, I say, untrue, because the only conceivable
basis of such an emotion to the creative mind
would be the creature's independence, and this
basis is utterly wanting, being swallowed up in his
sheer and ceaseless dependence. Thus, in order
that man really do anything either praiseworthy
or blameworthy in the Divine sight—in order, in
other words, that God Himself should charge us
with any of the good or evil which we with obdurate stupidity are for ever charging upon our-

selves—it would be necessary for Him first to for-
get His creative relation to us, and begin to look
upon us as essentially underived and independent
existences; which is absurd. I perfectly admit
that the truth, as reflected in fact, *seems* directly
otherwise. It actually does, and must, seem to
the sensuous understanding—the intelligence controlled by sense—that man is an absolute selfhood,
that is to say, that his affections and thoughts, far
from being an influx from spiritual association,
originate in himself exclusively, and hence leave
him properly chargeable with all the good and
evil issuing from such affections and thoughts.
The senses confined to the *seeming*, cannot help
bedevilling in this way our nascent scientific intellect. They recognize only what appears to them,
having no glimpse, however faint, of internal
realities; and hence they cannot but teach to every
one who seeks instruction at their hands, that the
actual is the only real, that the spiritual sphere, if
any such sphere exist, is only another natural,
governed by the same laws, and reproducing the
same phenomena. Thus they insinuate that our
physical finiteness—our visible insularity in time
and space—is a real and eternal truth. They teach
me that I am in all real or spiritual respects precisely what I am in natural or seeming ones, that
is to say, an utterly disconnected being, regarded
by God not as inseparably interwoven and united
with my kind, but as distinctly disunited with all

other existence, and governed by Him on strictly
private and special methods.

Hence it falls out that the dull and sombre walls
of our ecclesiastical Zion, and the less sombre but
flippant courts of our received philosophy, enclose
a far more organized hostility to spiritual Christianity than you will find in conventionally disreputable quarters. The scientific mind, like Pi-
late, "finds no evil" in the new Divine spirit
which is quickening the nations like life from the
dead: on the contrary, it dimly feels that the new
spirit is full of blessing for itself, and stands ready
to ask of it, "What is truth?" But the *soi-
disant* "regenerate" mind, we who think we see
—we who are not, like the vulgar herd, "accursed,
because they know not the law," but are in fact
sanctified by such knowledge, and actually rule the
world by its *prestige*—we feel our unrighteous
sway menaced by this tender and loving spirit,
and do, as the Jew did of old, everything we can
to ensure its endless triumph, by stupidly trying
to stifle and crush it. What the Jew did to Christ
in the flesh, was only a type, inexpressibly faint, of
what we Christians are daily doing to him in the
spirit. The Jew had never any power to harm
Jesus but by patronizing him. Had he done this,
had he espoused the Christian teaching and temper, Christ would have been bound indefinitely to
remain the mere Jew He was born, and there is
no saying accordingly how long Judaism might

have perpetuated itself, no longer indeed as a hurtful, but now as a beneficent, yoke upon the nations, nor consequently how long the Gentile mind might have failed to attain to the scientific sentiment of human equality, which yet is the exclusive basis of the Divine creation. So now, the only hindrance which our existing authorities in Church and State could offer to the new ideas, would be to patronize them, to lend them the furtherance of their adoption: for then the common mind of Christendom, which is very docile to good influences, would be so full of admiration and gratitude towards these old established and now undeniable stewards of God, that a new and worse idolatry, a new and more benumbing servitude of the human mind, would be sure to ensue, and a third advent of the Christ behoove to take place, in order to strike off the fetters forged by the preceding one. The new wine of Protestantism and Democracy—the spirit of an ever-advancing humanity—would seek in that case to confine itself evermore within the old established bottles of Church and State, within the purely symbolic dimensions of priest and king, and by dint of so seeking would be infallibly sure to turn vapid and lifeless, to tumble finally, in fact, into the condition of mere disreputable swipes, only fit to be poured out upon the ground, a scorn and avoidance to men and animals.

This, in literal verity, is the fatal sign about

European Christendom, that it has inherited in Christianity a soul altogether disproportionate to its meagre and inexpansive body. Protestantism is the actual limit of the Church's elasticity,—one strain more, and it snaps into Mormonism or other downright deviltry, which reasonable people will some day be forced to sweep bodily from the earth : and the State can go no further than Democracy without going into visible extinction. In fact, all astute priests and politicians have perceived for years past that Protestantism and Democracy are not so much expansions of the old symbolic institutions of Church and State, as actual disorganizations of them. They mark the old age of those institutions, their decline into the vale of years, preparatory to their final exit from the historic scene. Hence that prevalent movement of unbelief and despair among our upper classes in Church and State, which christens itself *Conservatism*, and which consists in seeking refuge from the onward Providence that governs the world, by flinging oneself into the arms of the stolidest civil and ecclesiastical despotisms, or in calling upon the mountains and rocks to crush one, by way of shielding one's eyes from the entrance of unwelcome light. How utterly absurd then to suppose our existing Christendom formally competent to embody the Divine spirit in humanity ! This spirit seeks the infinite expansion of human nature, seeks to lift the beggar from the dunghill and to set him among princes, simply because he is man, simply because he is a living form or image of God, and hence capable of an immortal conjunction with God. God is blessedly indifferent to the interests of every priesthood and every government under the sun, because He stands in an infinitely nearer attitude to man than these priesthoods and governments can any way conceive of as possible. They have not the slightest conception of God as the Lord, or of a Divine *natural* humanity, but on the contrary, maintain, under Christian names, the most inveterately Pagan conceptions of the Divine character. Take, for example, any reigning Pope or Emperor, and chase the Divine image through all the windings of his official heart down to its fundamental quality, and you will find it turn out some sheer personal will, some strenuous physical existence, reeling with the possession of mere wanton power, and odious from the exercise of every jealous revengeful and malignant disposition. It is high time that all the world confess themselves atheists with respect to this orthodox deity. It is high time that every disciple of Christ seize this obscene and skulking god of the nations by the beard with one hand, and with the other smite him between the eyes till he fall down and die. The famous M. Proudhon, who snaps his whip louder than any contemporary Frenchman, very much shocked his hypocritical generation a little

while since by crying *haro* upon this Gentile conception of God, or exclaiming against Deity thus
viewed as the true curse of human existence.
Proudhon's critics, who themselves are fond of
snapping their whips in the loudest possible way,
seem to have been disheartened by the tremendous
*eclât* of his performance, and are accordingly doing
what they can ever since to diminish· it, by representing it as a mere insincerity on Proudhon's part
—as a mere *annonce* to the travelling public that
here at last was a postillion capable of taking
them the shortest possible route to kingdom-come,
provided they would only commit themselves to
his audacious guidance.  I do not personally enjoy
the pleasure of M. Proudhon's acquaintance, but I
cannot help feeling very serious misgivings as to
the truth of this criticism.  His judgment strikes
me as on the whole a very Christian one.  I suppose that Proudhon would be as much disconcerted
to be called a Christian as those modest people of
whom we read in the Lord's similitude of the
kingdom of heaven, as replying to his beaming
smile of recognition for services rendered, " But
when saw we thee hungry, and fed thee; or thirsty,
and gave thee drink? "   Nevertheless I regard
Proudhon as at bottom, if not a-top—in heart, if
not in head—an excellent Christian. His intellect
has doubtless been sophisticated to some extent
by the dense and blinding obscurity which has
traditionally settled down upon the moral problem;

but he is obviously a man of the manliest make in
heart, and I do not see how any clear-sighted
reader of the four gospels, which turn all subsequent revolutionary literature into child's play,
can feel justified in denouncing him.  Of course I
mean the unadulterate gospels, not that bleached
and emasculate substitute which, under the name
of " evangelical religion," does its weekly best to
defame and deface God's image in our souls,
through the length and breadth of established
Church and State.  Evangelical religion as it is
called, *quasi lucus a non lucendo, quasi mons a non
movendo,* is such a religion as is fitly piped by the
east wind—a religion which cuts across the nerves
of the soul like a knife, which chills all the best
sympathies of the heart, and ends by freezing its
followers stiff in the shallows of their own selfishness.  It is of course not of this conventional
gospel that I speak, but of the unperverted gospel
of Christ, when I say that every intelligent reader
will be slow to condemn Proudhon, because
throughout his unskilful books he will yet not fail
to discern an unmistakeable flavour of that an-
cient and incomparable vintage.  Clearly, if Christianity makes any distinct pretension, it is to have
utterly exhausted natural religion; and natural
religion is the only thing with which the scientific
intellect of man has any quarrel.  Science revolts
at the idea of there being any essential limitation
of the human faculties, which nevertheless would

be inevitable if their vital source could be proved
to lay outside of human nature, or inhered as
natural religion affirms it to inhere, in a being
generically distinct from humanity, and *spatially*
separable from all its individual forms. Science
utterly revolts from the conception of a physical
or material Deity—a Deity cognizable to sense—
and triumphantly careers through the universe of
space, to chase from the human mind every vestige of so baleful and disheartening a conception.
But it is solely to Christianity that science owes
this emancipation. Christianity eternally explodes
the naturalistic conception of Deity as a being
essentially disproportionate to man, and therefore
inaccessible to human intelligence, by identifying
Him with conventionally the meanest and humblest of men, with a man who was so genuinely
humble and insignificant as actually to feel no
personality apart from the interests of universal
truth and justice, who had not spirit enough to be
angry at the grossest of personal insults, or to re-
sent the cruellest of personal wrongs; but, on the
contrary, habitually and patiently endured degradations which any rustic English pedagogue at the
present day would be parochially disowned for
submitting to for a moment, and which would
drive the most sonorous of your English bishops
to doubt the Divine existence, if he were even so
much as threatened with them. Yet He, adorable
man of men, bore unflinchingly on, nor ever

ceased to eat the bitter bread of humiliation, until
He had made his despised and suffering form the
adequate and ample temple of God, and so for
ever wedded the infinite Divine perfection to the
most familiar motions and appetites of our ordinary human nature.  Jesus vindicated his prophetic designation as above all men *" a man of
sorrows,"* because in the historic position to which
he found himself born, he was exposed on the one
side to the unmeasured influx of the Divine Love,
and on the other to the equally unmeasured influx
of every loathsome and hellish lust of personal
aggrandizement.  The literal form of Christ's pretension was profoundly diabolic.  View his personal
pretension as literally true and just, as having an
absolute basis, and you can imagine no more
flagrant dishonour to the Divine name.  To suppose that the universal Father of mankind cared
for the Jew one jot more than for the Gentile,
and that He cared for one Jew also more than for
another, actually intending to give both the former
and the latter an endless earthly dominion, was
manifestly to blacken the Divine character, and
pervert it to the inflammation of every diabolic
ambition.  And yet this was that literal form of
the Jewish hope to which Christ was born.  The
innocent babe opened his eyes upon mother and
father, brother and sister, neighbour and friend,
ruler and priest, stupidly agape at the marvels
which heralded his birth, and no doubt as his

intelligence dawned he lent a naturally compla-
cent ear to the promises of personal advancement
and glory they showered upon him.  He sucked
in the subtlest spiritual poison with every swallow
of his mother's milk, and his very religion bound
him, so far as human probabilities went, to become an unmitigated devil.  I find no trace of
any man in history being subject to the tempta-
tions that beset this truest of men.  I find no
trace of any other man who felt himself called
upon by the tenderest human love to loathe and
disavow the proud and yearning bosom that bore
him.  I find no other man in history whose profound reverence for infinite goodness and truth
drove him to renounce the religion of his fathers,
simply because that religion contemplated as its
issue his own supreme aggrandizement; and whose
profound love to man drove him to renounce every
obligation of patriotism, simply because these obligations were plainly coincident with the supremest
and subtlest inspirations of his own self-love.  No
doubt many a man has renounced his traditional
creed because it associated him with the obloquy
and contempt of his nation, or stood in the way
of his personal ambition; and so no doubt many
a man has abjured his country, because it dis-
claimed his title and ability to rule.  In short, a
thousand men can be found every day who do
both of these things from the instinct of self-love.
But the eternal peculiarity of the Christian fact

is, that Christ did them utterly without the aid
of that tremendous lever, actually while it was
undermining his force, and subjecting him to
ceaseless death. He discredited his paternal gods
simply because they were bent upon doing him
unlimited honour; and shrank from kindred and
countrymen, only because they were intent upon
rendering him unparalleled gratitude and benediction. What a mere obscenity every great name
in history confesses itself beside this spotless Ju-
dean youth, who in the thickest night of time,—
unhelped by priest or ruler, by friend or neighbour, by father or mother, by brother or sister,
helped, in fact, if we may so consider it, only by
the dim expectant sympathy of that hungry rabble
of harlots and outcasts who furnished His inglo-
rious retinue, and still further drew upon Him
the ferocious scorn of all that was devout, and
honourable and powerful in His nation,—yet let
in eternal daylight upon the soul, by steadfastly
expanding in his private spirit to the dimensions
of universal humanity, so bringing, for the first
time in history, the finite human bosom into perfect experimental accord with the infinite Divine
Love. For my part I am free to declare that I
find the conception of any Divinity superior to
this radiant human form, inexpressibly treasonable
to my own manhood. In fact, I do not hesitate
to say that I find the orthodox and popular conception of Deity to be in the comparison a mere

odious stench in the nostrils, against which I here
indite my exuberant and eternal protest. I shall
always cherish the most hearty and cheerful
atheism towards every deity but him who has
illustrated my own nature with such resplendent
power, as to make me feel that man henceforth is
the only name of honour, and that any God out of
the strictest human proportions, any God with
essentially disproportionate aims and ends to man,
is an unmixed superfluity and nuisance. In short,
I worship the LORD alone, the God-MAN, that peerless and perfect soul whose unswerving innocence
and sweetness gathered up the infinite forces of
Deity as wheat is gathered up in a sheaf, and for
ever linked them with the natural life of man,
with every commonest lineament of human nature,
so that we are not only authorized henceforth to
view the human spirit as inwardly refined from all
grossness, which is pride or selfishness, and instinct with universal love and humility, but also
to regard the human body itself as the only visible
shrine of God, as the destined temple of all lustrous health and beauty, the native home of every
chaste, and generous, and magnanimous affection.
I take it that every man of sense and feeling will
infallibly join in this ennobling worship. I take
it that all atheism and scepticism are inwardly
fragrant with this devout incense, that to the loving and knowing heart of God they have never
been anything else than a negative but most sin-

L

cere form of the vital worship I here avow. It is,
indeed, obvious that Proudhon's manly revolt con-
templates only that old Pagan conception of the
Godhead which Christianity exhausts, but which
nominally Christian priests and kings, for their
own private unloving ends, still continue diligently
to *exploit*. Against this lurid power—half-peda-
gogue, half-policeman, but wholly imbecile in both
aspects—I, too, raise my gleeful fist, I lift my
scornful foot, I invoke the self-respect of my children, I arouse their generous indignation, I instruct their nascent philanthropy; because I know
that he spiritually departed this life long centuries
ago, and that it is only his grim unburied corpse
which still poisons the popular air.

But now, although I say all this *ex animo*, do
not, I beseech you, regard me as echoing, in any
measure, the tedious cant of orthodoxy. If I
heartily detest anything it is our existing Christian
Judaism (the exact antitype of what the four
gospels describe to us in type), with its wrangling
regiments of spiritual old-clothesmen diligently dividing the empty garments of Truth among themselves, and hawking the dislocated fragments about
as if they were the immortal substance itself. As
I have already said, the letter of Christianity constitutes only the seeming or phenomenal aspect of
Divine Truth, the semblance which it puts on to a
sensuous intelligence, an intelligence not *inwardly*
enlightened. It gives us very much the same un-

worthy impression of the Divine Truth as a child
would form of its father's tenderness who should
see that tenderness only in negative exercise, that
is, incessantly employed in restraining its natural
evils, correcting its fallacious judgments, in short,
educating and disciplining it into true human proportions out of its native wilfulness and conceit.
In a word, the letter of truth is *ipso facto* bound
to prove a purely negative and symbolic utterance
of its substance or spirit. This obligation flows
from the great law which makes the natural, in all
cases, an inverse expression of the spiritual, or
renders the body the bounded home and continent
of the boundless soul. My inmost soul, or life, is
the infinite God, is perfect goodness and wisdom :
but manifestly, unless I had some natural limitations, some finite continent (so to speak) separating *me* from *you* and every other body, I should
never appropriate this soul, or life, should never
be able to feel it and name it *me, my* self, should
be destitute, in a word, of conscious existence.
But now this bodily or finite me, which seems the
most incontestable of facts, is nevertheless the exact
inversion and denial of the infinite truth. It is
the imprisonment of the infinite love and wisdom
in the purely specious shackles of space and time.
The spiritual truth is, that there is but one life,
God, and that He alone lives in us : but this would
be death to feel, though it is life to believe it ;
because if we *sensibly felt*, as well as rationally

L 2

believed, that God alone lived, it is obvious that
we ourselves should become instantly converted
into stocks and stones, into the breathless images
of unbreathing men.  His superb mercy, above all
things, provides therefore that we shall never *feel*
this truth to all eternity, that however we may
reflectively think and believe in the premises, it
shall yet always sensibly seem to us that life is
disunited, is infinitely various, and that we are its
absolute proprietors.  In short, the Divine Providence perpetually endows us with selfhood, perpetually ensures that we shall *feel* the finite me to
be the most indisputable of realities.  But now, if
He left us there, mere creatures of sense: if He
did not go on to educate us out of our purely
physical consciousness by the inspirations of conscience, by developing in us the most passionate
social relations, so linking us with parent and
friend, lover and neighbour, fellow-countryman
and fellow-man, till at last our existence became
widened to the dimensions of universal humanity,
we should never discern the spiritual truth of the
case, but remain under the dominion of mere
natural appearances, the victims of the silliest pride
and self-complacency, to the end of the chapter.

Now the letter of Revelation bears a precisely
analogous relation to *its* spirit.  It furnishes a
purely negative index to its own substantial contents, because it is addressed to an unspiritual intelligence, and hence is bound to mask itself in

such coarse features as shall be sure to conciliate,
or at all events not revolt, that intelligence.  But
if we hereupon stupidly insist upon confounding
letter and spirit, if we insist upon the former not
as a purely representative or symbolic, but as a
direct and adequate expression of the latter, we
shall completely miss the true scope of all Divine
revelation, and remain mere spiritual embryos and
abortions to all eternity.  Spiritual substance, as
Swedenborg shews, has nothing in common with
time, space, and person.  The literal Christian
facts in his view constitute neither more nor less
than a revelation, within the sphere of sense, of a
life in man which profoundly subtends his senses,
but which yet could never come to consciousness
in him save in the very same way that all supersensuous ideas come to consciousness, that is, by
means of some sensible revelation or imagery,
serving as a mould to give them development.
All mankind, for example, have the idea of God as
the infinitude or perfection of character, of personality.  But we could never recognize character
or personality in God or man without the mould
which our moral experience supplies to that perception.  My moral experience tells me that justice
is good and injustice is evil, that he who injures
his neighbour is an evil man, and he who refrains
from injuring him a good man.  Now these moral
judgments serve simply as a mould or body to our
spiritual perceptions, and being as such mould or

body the exact inversion of what is moulded or
embodied in them, they have obviously no more
right to control our spiritual perceptions than an
egg has to control the chicken, than the foundation
of a house has to control the superstructure, than
the kitchen has to control the drawing-room, than
the stream has to control the fountain.* But they
are, as I have said, an invaluable and indispensable
basis and servant of those perceptions. My moral
judgments serve, in fact, as a rude but genial
mother-earth for the outgrowth of my spiritual in-

* For example, if we should pronounce a man spiritually good
simply because he was morally good, or spiritually evil simply
because he was morally evil, we should be guilty of gross absurdity, because, in reality, no human being has the slightest underived moral power, and it is only underived power whose activity
confers responsibility. I have no power to injure my neighbour
which is not derived to me from hell, or evil association, nor any
power to refrain from injuring him which is not derived to me
from heaven, or good association; and I am not spiritually chargeable, therefore, with either my moral good or evil, but only naturally chargeable with it. They are both alike a mere natural
inheritance, the legacy of my past ancestry. No matter how diligently soever I may work this inheritance, I can do no more at
best than associate myself with heaven or hell. I may have all the
moral virtue that has ever inflamed human pride, and I shall not
be one whit nearer the fountain of life. I may have all the moral
infirmity that has ever quickened human despair, and I shall be no
whit more remote from it. For that life surrounds human nature,
as the waters surround the earth, bathing equally both its contrasted poles; and we might, with precisely the same propriety,
deny to the ocean its measured tides, its alternate ebb and flow, as
to the Divine life in humanity its perpetual sportive interchange
and conjugation of brotherly love and self-love.

telligence. Unless I first felt in myself a moral
personality, constituted of the exact equilibrium
of good and evil, or heaven and hell, I should lack
the fundamental germ of that subsequent spiritual
conception of myself, which presents the subjection
of evil to good, and of both to the Divine. My
true life is a spontaneous one, a life of taste or attraction, a life of freedom, growing out of a complete reconciliation of self-love with brotherly love,
the true man never seeking his own ends but by
assiduously promoting those of universal man.
But clearly I should never be able to grasp or even
discern this perfect life, save by the contrast of a
previous unreal or enslaved one. If I were not first
delivered over by conscience to the experience of
death in myself as finitely organized, as vivified
by nature and custom, I could never have realized,
nor even aspired to realize, that perfect life or
righteousness which inheres in myself as Divinely
organized, as vivified by infinite love and wisdom.
Thus, as I say, my moral experience serves no
higher end than to incarnate, or give body to, my
spiritual life. In short, the moral man, good and
evil, is but the inversion or shadow—is but the
rude decaying germ or egg—is but the perishable
natural body—of the imperishable spiritual man,
who is Divinely or immaculately good, good without the slightest antagonism of evil.

Now, I repeat, that the letter of Revelation ob-
serves precisely this same servile relation towards

its proper spiritual substance. The letter is but
the perishable husk of the imperishable spirit.
The literal dogma, for example, of Christ's divinity,
is wholly unintelligible in heaven, because, as
Swedenborg shews, heavenly thought is never determined to person, but only to the things represented by person. In short, the spiritual contents
of the dogma alone are apprehended in heaven,
and these are that human nature itself is Divinely
vivified, is the adequate and ample abode of perfect love and wisdom. The literal dogma is the
needful egg (so to speak), is the indispensable preliminary basis of our subsequent scientific acknowledgment of the exclusive Divinity of our natural
origin. Had we not been taught, traditionally, to
regard this most humble and abject partaker of
our nature as Divine, as perfectly united with
infinite power and goodness, spite of his total
destitution of whatsoever men are wont to admire
in character and manners, of everything that gets
itself eulogized, for example, in our great flaunting
and mendacious newspapers, our present scientific
assurance that human nature itself is Divinely
quickened, could never have even germinated.
The Christian truth is the sole ground of the difference between the scientific mind of the race and
the unscientific mind, between the public conscience, for example, of Christendom and that of
Mahommedanism. Take away the traditional
Christian dogma from our annals, and the long

expansion it has lent to the human faculties, and science would still be groping in the sublimated mud of alchemy and astrology, or perhaps gravely discussing, along with the theologian and philosopher, the momentous question, whether or not God was identical with the contents of a certain sanctified bread-basket.

Remember, then, that the literal dogma is in every case only a needful platform of the super-sensuous truth, bearing a directly inverse relation to its spirit, such as your image in a mirror bears to yourself, or the outside of a glove to its inside. Thus the Divine incarnation, spiritually viewed, is a universal truth, having no more validity to one man's experience than to another's. This transcendent truth was indeed completely *revealed* in the Christ, but you would not confound the external revelation of a truth with its interior substance, any more than you would confound a negative with its positive, body with soul, or your transient shadow in the looking-glass with your living self. In fact, you are inexorably forbidden to do so, as we have already seen, by the circumstance that the letter of revelation, in virtue of the baseness of the intelligence to which it is addressed, has never any pretension to be worthy of its spiritual contents, except as the body is worthy of the soul, the shadow of its substance, the servant of his master, that is by negatively reflecting it. If the servant were a positive reflection of his lord,

L 3

the shadow a positive reflection of its substance,
the body a positive reflection of the soul, there
would be no such thing as choosing between servant and lord, between shadow and substance,
between body and soul. In short, we should live
in a highly ridiculous world, in which all the
needs of the human understanding had been wan-
tonly violated. Analogically, then, the letter of
revelation, by virtue of the limited intelligence to
which it is addressed, is bound to obscure and
falsify, to some extent, its own spiritual contents,
just as the squint eyes, the crooked back, or inverted feet I have inherited from my past ancestry,
obscure my spiritual form, my substantial contents, or as your image in a glass being addressed
to your bodily, not your mental eye, falsifies your
proper self-consciousness, turning what your mental eye pronounces your right-hand into your left,
and so forth. It will not do, therefore, whatever
the bare face of revelation declares, *spiritually* to
assert a limitary incarnation of Deity, such an incarnation as not only *apparently* but *really* restricts
Him to specific times, places, and persons. Because, if we do thus, we shall infallibly stifle the
true scientific and spiritual conception which incessantly postulates His infinitude, that is, His
complete exemption from these finite bonds.

Let us fully accept then the literal Christian
dogma, but only as the indispensable basis of that
sovereign spiritual verity, which lifts the Divine

incarnation out of the realm of mere sensuous
appearances—out of the limitations imposed by
our natural stupidity—into a strictly universal
truth, or one which is illustrated in every individual bosom of the race. The spiritual substance
embodied in the literal Christian verity, is, that
God vivifies man *naturally* no less than spiritually.
It imports—no longer that this, that, and the
other person becomes conjoined with God by his
proper spiritual fermentation and ripening, but—
that human nature itself, *by its own distinctive*
*process of fermentation and ripening denominated*
*history,* becomes henceforth eternally conjoined
with the same Divine perfection. In fact, the
Christian truth implies that all our private re-
generative experiences have been only so many
faint and feeble *primitiæ* of this grand public
operation of God, only so many timid and star-
veling rills of this affluent Divine fountain in the
very bosom of the race itself. This is the exact
meaning of history, a process of spiritual fermentation and refining within the public or associated
consciousness of man, or what is the same thing,
the regeneration of our very nature. It means
the development of a selfhood in man adequate to
image the creative infinitude, and therefore scientifically fit to avouch the Divine creation. It
means the gradual coming to consciousness on the
part of the race, of its intimate and eternal alliance with all divine power and beauty; in short,

the evolution of a Divine NATURAL manhood. Thus, as every true biography vindicates its claim to be written, only by relating how some private person, from being the abject offspring of his parents, became by God's inward nourishment a living soul or selfhood, capable of rising eternally away from his earthly nest, and forgetting on occasion every rudimental natural tie: so all veritable history busies itself with relating how that public person whom we denominate human nature becomes lifted by God's secret and ceaseless inspiration out of the abject mud of space and time, out of its purely mineral, vegetable, and animal anchorage, into the conscious fellowship of infinite goodness, and the consequent eternal supremacy of all inferior natures. Man has both a common or public personality and a private one: there is both a mind of the race and an individual mind: and the perfected scope of the Divine Providence or the consummation of human history, is the due CO-ORDINATION OF THESE DIVERGENT ELEMENTS, *the interior or superior place accruing by every title to the individual or feminine element.* But it is notorious that man has never intelligently seconded the divine purpose herein. On the contrary he has always done his most pompous best to resist it. His most accredited theologies and philosophies have diligently taught him, by sensual instigation, that Eve was essentially subject to Adam, that is, that the private or individual force in man was rightfully secondary and servile to the common or public force: and hence it is the invariable lot of these theologies and philosophies to find themselves disowned by the advance of history, which is the growth of man's scientific insight. History quietly antiquates and paralyzes every creed, sacred or secular, which defames the human soul by representing it as *freely* alienating itself from God: because the sole beatific function of history is to prove such alienation impossible, save under conditions of servitude, when the mind is a prey to the tyranny of ignorance and superstition.

The march of history incessantly vindicates the rightful primacy of the affections, or what is the same thing, incessantly quickens the spontaneous force in us, by depressing our voluntary or moral force.* The moral life of man is a phenomenon of our scientific immaturity. It grows out of our appropriating to ourselves the good or the evil we do, instead of ascribing it exclusively to the hereditary influx of good and evil spirits, and hence feeling no more responsibility for it, no more sense of merit or demerit in regard to it, than we should feel in regard to a fair or muddy complexion, to a sunny or sombre natural disposition. So long as we continue stupidly to munch this pestilent fruit, it is of course inevitable that we find ourselves excluded from the Tree of Life. I say "of course," because manifestly all the

* See *Appendix D*.

while we go on to appropriate this strictly influent
good and evil, we cannot help attributing to ourselves a purely simplistic or differential selfhood,
so remaining utterly blind to the great scientific
truth of our unitary or composite existence: and,
coming before God in that miserly plight, in that
lean and penurious condition, the voice of the
Divine mercy towards us is bound to shroud itself
in tones of despair, only faintly relieved by distant hope. For God sees us only in the intensest
unity with our kind, only in indissoluble solidarity
with every other individual of the race; and consequently, whilst we view ourselves as independently constituted, as related to Him by our own
absolute merit or demerit, irrespectively of our
connexion with the race, we must necessarily be
full either of egotistic pride or equally egotistic
despair, and in both cases alike can hardly help
proving an extremely unsatisfactory spectacle to
Him. What should we think of an eye or a hand
that deemed itself related to the light and air by
itself, and independently of its connexion with
the body? Why, obviously, that it was diseased
and ready to perish. Well, the infinite wisdom
makes precisely that judgment of us, when we
fancy ourselves righteous or unrighteous in our
own right, and apart from our unity with our
kind. There is no pretension more insufferably
arrogant in the Divine sight than that of any
merely individual ability to keep the Divine law.

I am persuaded that I never cut a more con-
temptible figure in the Divine estimation, than
when I suppose myself capable of refraining from
stealing my neighbour's purse, or seducing my
neighbour's wife, by some private force of my
own, and independently of angelic association, or
of the help I derive from my connexion with the
race. And I presume on the other hand that there
is no attitude of mind more intrinsically respectable in the Divine sight, more cordially delightful
to the Divine mind, than that which should exhibit
the thief or adulterer totally indifferent to the
unrighteousness which is conventionally charged
upon his private character, while he calmly referred
all the evil of his conduct to the wholly unscientific aspect of our social relations, to the shock-
ingly imperfect way in which the sentiment of
human equality or fellowship is yet organized in
institutions. God hates nothing on earth but
kings and priests: that is to say, never the veritable human persons that are hereditarily or traditionally swaddled in those effete offices, but the
offices or institutions themselves so named: because they are the only things which now obstruct
the Divine kingdom upon earth, by hindering the
scientific organization of human fellowship. And
whatsoever hinders that, His perfect love to mankind bids Him hate, bids Him hand over to speedy
and remorseless destruction. I am for my own
part neither a thief nor an adulterer, but I could

almost long to be both one and the other after the
most flagrant type, that thus I might fling back
with exquisite scorn the imputation of unrighteousness wherewith society would seek in that case
to cover me—or rather, that thus I might drink
in with keener relish the profound conviction
which all history, which all science, brings home
to me, namely, that in my real, my spiritual,
private, and God-given self I am wholly incapable
of evil either in affection, thought, or action, and
that it is therefore only in my *quasi*, my conventional, public, and man-given self, that I ever find
myself incurring such liability. Thus I would never
seek to hide, but rather to make conspicuous, all
the iniquity charged upon me: only I would insist upon its being an iniquity which attached to
me, not as disconnected with other men, but as
intimately blent and bound up with priest and
king, with teacher and ruler, with every devout
and honourable person in short, who is officially
interested in maintaining the existing infirm organization of human society or fellowship.*

But I can no longer afford these digressions,
which after all are no digressions, except to a hur-
ried observation. My space warns me to come
rapidly to a close. I have just said that the progress of history in depressing the moral vigour of
the race, operates an incessant elevation of its
spontaneous force. This result ensues by virtue

* See *Appendix E.*

of the same law which in the physical sphere limits
the menstrual flux by the phenomena of conception
and gestation.  For morality is exactly the same
phenomenon in the spiritual sphere, or the life of
the race, which menstruation is in the natural
sphere, or the life of woman, that is to say, it is
a process of elimination or purification; and it
operates precisely the same uses, that is to say, it
abates the natural pride and vigour of the heart,
and so disposes it to conceive and bring forth
spiritual fruit.*  The end of conscience is to pu-

* Recent physiological researches go to shew that the menstrual flux signalizes the spontaneous maturation of the ovum and
its consequent separation from the ovary and descent into the
uterus, for the purpose of impregnation.  At all events, it seems
to be clearly established, that conception ordinarily takes place
just before or just after menstruation, and is very rare at other
times; so that we may fairly infer a very close connexion between
the two.  But the science of correspondences, which is the only
Divine science, because it is the science of the very sciences themselves—turning the sandy wilderness of disconnected facts which
they present to us into the unity of a blooming garden—dissipates
all doubt as to the function of menstruation, by turning it into a
strict analogon and ultimate of that great spiritual ordeal of purification which we denominate conscience.  The aim of menstruation is purification, is such a vastation of the native grossness of
- the body, as disposes it to conception and prolification.  Consequently, until menstruation begins, conception is impossible, and
it is equally impossible after menstruation has ceased.  Then,
again, woman alone menstruates, because she is a natural form or
representation of the selfhood in man, or of that thing which is
eventually to ally him with God by redeeming him from animality:
and it is only the selfhood as still unconscious of its function and
beguiled by the senses, or the fallacious shows of things, that con-

rify, and so prepare the soul for immortal conjunction with God : and purification means that gradual depletion of the natural selfhood or proprium which constitutes all that is valuable in our historic experience. I know very well that morality is not popularly supposed to play this subordinate part in human affairs. Every consistent churchman and statesman will revolt at my assigning it this strictly ministerial office, this purely solvent or transitional efficacy. It constitutes in fact the still invincible strength of hell on earth, that morality is everywhere looked upon as having a properly magisterial authority, as furnishing the indisputable Divine breath of our spiritual life. You might with equal propriety look upon physics as furnishing not merely the outward condition—the necessary platform or base—of our moral life, but its inward

science seeks to purify. The mere Adamic or animal life is innocent enough in all the range of its passions and appetites, and consequently invites no purgation. It is only the Divinely-given selfhood of man, which, owing to its ignorance and inexperience of its true source is for a long time unworthily duped by the senses, and so subjected to the Adamic or bodily rule, that demands chastisement. Hence it is that woman alone, being the true analogon of this selfhood, menstruates, and so becomes physically qualified for maternity. If you wish any light upon the physiological question here adverted to, you may consult a careful and conscientious work of M. POUCHET, entitled *Theorie de l'Ovulation Spontanée*, and a little book of RACIBORSKI on the same subject, which I have also read with interest, but whose title I do not now recall. The supplement to Baly's translation of *Müller's Physiology* furnishes a good abstract of *all* the literature of the topic.

substance also. The wrong done to truth in either case is precisely the same. In fact the peacock who parades his lustrous plumage to captivate our admiration, is only a sensible type of that subtler foppery, of that more harmful pharisaism, which confounds moral distinctions with spiritual, or supposes a man divinely vivified not by what unites him with other men, but only by what separates him from them. Hell has no profounder root than this.* The entire diabolic *nisus* in humanity

---

* Hell is nothing but the gradual sloughing-off or separation in the angelic mind of self-love from charity, which separation is necessitated so long as the Divine life in nature is practically inchoate. The Divine NATURAL man of course comprehends in his own person both heaven and hell, and reconciles them equally to the Divine good: or if a difference be insisted on, he makes the latter even *more* tributary than the former to that good. But until that achievement becomes so far consummated in interior realms of creation as to be avouched to our natural consciousness by the plenary diffusion of the Holy Spirit, the promised Comforter, which is the truly scientific spirit of human fellowship or equality, heaven and hell remain at war, and the angel grows an angel only by the spiritual elimination and precipitation of what in him is hereditarily diabolic. Thus angelic existence confesses itself undivine by all the bulk of those various hells, which it voids upon the universe in the process of asserting itself. The hells are only so much incomparable Divine force spiritually disowned by the angel, turned to waste by his sheer incapacity freely to image God, that is, to do good spontaneously. Indeed the heavens had long ere now been swamped and stifled in their own proper ordure, had not the Divine Wisdom known how to utilize the lowest hells (even as the skilful husbandman knows how to utilize his festering heaps of manure), by transforming them into the substance of a new and more glorious manhood.

dates indeed from this grossly fallacious estimate
of truth. It is the infirmity of the unscientific
mind, of the understanding enlightened only by
sense to confound nature with spirit, fact with
truth : to mistake the actual for the real or seeming for being.   Thus, inasmuch as I sensibly *appear* to be an absolute existence, or to have a selfhood utterly distinct from and independent of
angel and devil, my unpractised reason is incontinently beguiled to conclude that such is really the
case, and hastens to confirm the shallow fallacy by
zealously affiliating to my spiritual *self* all the
good and evil which hereditarily influence my
*nature :* so filling me in spirit with an odious self-
conceit or an equally odious self-distrust, which
both alike engender hell in me, because they both
alike exclude that bosom-peace which makes the
immortal substance of every bliss known to heaven.
Hell has no root but human pride, and the earth
by which that root thrives would be instantly dissolved, were we manfully to cease " eating of the
tree of knowledge of good and evil :" that is, cease
attributing to ourselves the moral traits which flow
solely from our hereditary connexion with heaven
and hell, by appropriating those superior spiritual
qualities which come to us from God alone, and
which presuppose the complete reconciliation of
hell with heaven.  At all events such has been the
undeniable drift of history.  That great institution
which we call the *Church*, has had no other aim,

from the beginning of history, than to depress the
moral consciousness of man, or shame him out of
pride and boastfulness, by exalting his æsthetic
consciousness, or making him feel that he is what
he is, not by virtue of any difference between him
and other men, but only by virtue of his intense
unity with them.  Religion, revelation, has had
no diviner office than to convince mankind that
their highest virtue, morally regarded, all that
virtue which exalts one man above another in
social estimation, and so enacts the reign of hell
on earth, is filthy rags in God's sight; because,
when men are once 'persuaded of this, they will
gladly accept the righteousness which is revealed
to them from God out of heaven, and which needs
no inauguration but that which is afforded it by
the scientific recognition of the great truth of
human society or fellowship.  Of course all that
is sentimental in you will howl at this assertion,
as feeling the very breath of its nostrils invaded :
for sentimentalism enjoys a purely outward and
osculatory dalliance with truth, and if deeper relations be insisted on, nothing is left it but to go
forth with the most sentimental of the apostles
and hang itself.  But I speak advisedly after years
of patient inquiry, and no amount of clamour can
affect my conviction that the truth I here allege
constitutes the adamantine basis of creation.

Yours truly,

————.

# APPENDIX.

_____

## *A.*—p. 138.

I quote here a few pages from a previous book of mine
now out of print, entitled *Lectures and Miscellanies* :—

" When I speak of the influence of ghostly communications upon ' weak-minded persons,' I mean persons
who, like myself, have been educated in sheerly erroneous
views of individual responsibility. After my religious
life dawned, my day was turned into hideous and unre-
lieved night by tacit ghostly visitations. I not merely
repented myself, as one of my theological teachers deemed
it incumbent on me, of Adam's transgression, but every
dubious transaction I had been engaged in from my youth
up, no matter how insignificant soever, crept forth from
its oblivious slime to paralyze my soul with threats of
God's judgment. So paltry an incident of my youth as
the throwing snow-balls, and that effectually too, at a
younger brother in order to prevent his following me at
play, had power, I recollect, to keep me awake all night,
bedewing my pillow with tears, and beseeching God to
grant me forgiveness. By dint of indefatigable prayer

and other ritual observance, I managed indeed to stave off actual despair from the beginning; and juster views of the divine character obtained from the New Testament, gradually illumined my very dense understanding, and gave me comparative peace. But I had no satisfactory glimpse of the source of all the infernal jugglery I had undergone till I learned from Swedenborg, that it proceeds from certain ghostly busy-bodies intent upon reducing the human mind to their subjection, and availing themselves for this purpose of every sensuous and fallacious idea we entertain of God, and of every disagreeable memory we retain of our own conduct.

" I call this information 'satisfactory,' because it accorded with my own observation. The suffering I underwent confessed itself an infliction, an imposition. I writhed under it as you have seen a beast writhe under a burden too heavy for him to lift, yet not quite heavy enough to crush him out of life. For I could not accept the imputation borne in upon me, that I was really chargeable with the guilt of any of these remembered iniquities. I of course did not deny an external or instrumental connexion with them; I did not deny that my *hand* had incurred defilement, but with my total heart and mind I resisted any closer affiliation. In reference, for example, to the trivial incident above specified, even while weeping scalding tears over its remembrance, I could not but be conscious of a present tenderness toward the imaginary sufferer, so cordial and so profuse as totally to acquit my inner or vital self of any complicity in the premises. Hence I had little doubt that the fact might be as Swedenborg alleged, and that I had been all along nourishing, by means of certain falsities in my intellect, a brood of ghostly loafers who had at last very nearly turned me out of house and home.

"It is not uncommon to hear the canting remark, that the world would be better off if men had a little more of the suffering in question. I have no objection to every man understanding the evil of his doings. On the contrary, I wish that every one might clearly discern his habitual iniquities, because until this discernment takes place, we shall not be in haste to put them away from us. But we shall never be able truly to confess them with the heart, *so long as we believe ourselves the source of them*— so long as we believe in our individual responsibility for them. The first step toward my acknowledging the evil of my doings, is my perception of its being a foreign influx or importation. If I view it as indigenous, of course I cannot deem it evil, for you would not have the same soil which brings forth the fruit condemn it also, would you? No man is wiser than himself. How therefore can you expect any one to acknowledge an evil in his conduct, unless you tacitly attribute to him an inward or essential superiority to that evil? If the evil come strictly from himself or within, if it do not proceed merely from defective culture, but grow out of the very substance of his individuality, then you simply insult him by asking him to repent it, or turn away from it. Would you ask a crab-apple stock to produce peaches, or a bramble-bush to bring forth grapes? Why then stultify yourself by expecting the peaceable fruits of righteousness from those whom at the same time you teach to regard themselves as the sources of their sin?

"I do not read that John the Baptist, who was reckoned a pattern revivalist, ever taught people to get up a

M

spiritual fidget, by way of qualifying themselves for the acknowledgment of the coming divine man. I read that he simply told each man to repent him of, or forsake, the evils incident to his proper vocation, the manifest patent evils which all men recognized and suffered from, and so stand prepared to do the will of the coming teacher. The attempt to fasten the authorship or responsibility of these offences upon the individual soul, and to establish the subject's metaphysical property in them, he left to the bloodhound sagacity of our modern theologians. It may be very grand and lofty in these perfunctory gentlemen to discourse upon the depth of human depravity, and so forth, but I have no hesitation in saying that the man who would really aggravate the self-condemnation of another, or intensify instead of moderate his conviction of personal defilement, no matter on what pretext soever of benevolence, is either himself grossly inexperienced in this horrid category of suffering, or else, may boast a heart harder than the nether mill-stone. He may have had what he calls troubles of conscience, but they have simply been got up for an occasion, got up with a view to his passing muster with his sect, or boasting an orthodox religious experience. An immense deal of this spiritual dilettantism exists in the world. The mere outside foppery we see in Broadway is as the fragrance of fresh hay in comparison with it.

"No one can object to another kindly pointing out any of his discernible evils of life, because every man feels it due to his manhood to rid it of all impediment. But clearly this is a very different thing from the endeavour to affix guilt to the soul. I know nothing so profoundly diabolic as this endeavour, whencesoever it may be exerted, from the pulpit or the closet, and for whatsoever ends, whether conventionally sacred or profane. To aim at making a poor wretch feel, that while simply obeying some dictate of nature, or perhaps some prompting of wounded passion, he has mortally affronted the very source of his life—that he *even has it in his power* to affront it—is a wickedness beside which, it appears to me, most of our burglaries and murders seem commonplace and tender. It is *spiritual* murder, murder not of the mere perishing body, but of the imperishable soul. And the man who is guilty of it, should be put to the penalty of silence for the remainder of his days, or at least until he proves himself better instructed. He very probably has a bosom full of parental tenderness, even while he is making so deadly an assault upon you in the name of his God, and would sooner renounce his own life than cherish a vindictive temper towards his dependent offspring. In which case of course, he is vastly more worshipful than the fetish he serves.

"But you say that this man does not leave you hopeless, that even while charging guilt upon you, he points you to the all-sufficient remedy for it. Alas! this apology proceeds upon the notion that a man's relation to God is merely physical or external, and that consequently provided he escapes a literal scourging from the divine hand, his aspirations are satisfied. Let every one speak for himself here. For my part, I am free to say, that I should be far more profoundly horrified by the idea of my *capacity* to offend God—even though I should *never actually do it*—than I should be by a fear of all the literal scourgings possible to be inflicted upon me, by all the self-styled deities of the universe. A deity who has it either in his hand or his heart, to inflict a wound upon any form of sensitive existence, is a deity of decidedly puerile and disreputable pattern. He is no deity for cultivated men and women. A deity whose prestige is chiefly muscular, arising from his imagined ability to inflict suffering, may still serve the needs of the Bushman, or the Choctaw, or our own rowdies : but to those in whom God's life has dawned however faintly, and whose souls accordingly are evermore consecrated to beauty, he is an unmitigated abomination. For a person of this quality knows no outward relations to God, no such relations as are contemplated or provided for by your mere pugilistic deity. God is his inmost life, without whom in fact he does not live : God is his vital selfhood, without whom indeed he is not himself : to talk therefore of enmity between him and God, is to talk of dividing him asunder, is to talk of separating his form from his substance, his existence from his being.

" I distrust accordingly these ghostly busy-bodies, who address our outward ear with gossip of the other world. They first arrest our attention by talk of those we have loved : they gradually inflame our ascetic ambition, our ambition after spiritual distinction : and finally, having got a secure hold, who knows through what pools of voluntary filth and degradation they may drag us ? I of course believe that spiritual help is incessantly enjoyed by man, but then it is a help directed exclusively to his affections and thoughts, not to his timorous and servile senses. The spiritual succour which comes in the way of quickening my intellect and affections, I am grateful for. It does not degrade me. It aggrandizes me, and makes my life more free. But that which comes in the form of outward and personal dictation, is an insult to my manhood, and in so far as it is tolerated, undermines it. It makes my will servile to a foreign inspiration, discharges my soul of its inherent divinity, and finally leaves me a dismal wreck, high and dry on the sands of superstition. It reduces me in fact below the level of the brute, for the brute has a certain reflected or colonial manhood, which disqualifies him for the tacit endurance of oppression. I am not speaking of impossibilities. We have all heard of tender and devout persons, who having through some foolish asceticism, or other accidental cause, come under the influence of this attenuated despotism, have at last got back to their own firesides, so spent with suffering, so lacerated to the very core, as to be fit—when not aroused to an indignant and manly reaction—only for the soothing shelter of the grave.

"On the whole I am led to regard these so-called 'spirits' rather as so many vermin revealing themselves in the tumble-down walls of our old theological hostelry, than as any very saintly and sweet persons, whose acquaintance it were edifying or even comfortable to make. I hope their pale activity—their bloodless and ghastly vivacity—may do indirect good by promoting a general disgust for the abject personal gossip which they deal out to us, and which has so long furnished the staple spiritual commodity of the old theology. But I vehemently discredit the prospect of any positive good. Man's true good never comes from without him, but only from the depths of divinity within him, and whatever tends to divert his attention from this truth, and fix it on Mahommedan paradises, and salvation through electricity, claims his most vindictive anathema. Above all, a spi-

ritual life which feels itself depleted by the diligent prose-
cution of the natural one, which is actually interested to
invade the latter, and persuade good sound flesh and
blood to barter its savoury cakes and ale for trite and
faded sentimentalities, is a life which every reasonable
person may safely scout as unworthy his aspiration.

"The mere personal gossip these ghostly gents remit
to us, proves of what a flimsy and gossamer quality they
themselves are, and how feeble a grasp they have yet
achieved of life. I am told that a communication was
lately received from Tom Paine and Ethan Allen, saying
that they were boarding at a hotel kept by John Bunyan,
and I can readily fancy the shaking of sides, and the rich
asthmatic wheeze, wherewith that communication was
launched by the inveterate wags who projected it. But
we are also told very seriously, that the apostle Paul and
other distinguished persons, have each a chosen medium
in our neighbourhood, on whom to dump his particular
wisdom, and so establish a depôt for that commodity.
And I learn besides that Dr. Franklin, Dr. Channing,
and several other well-behaved persons, are turning out
mere incontinent busy-bodies, and instead of attending
to their own affairs, have actually turned round again in
the endeavour to instruct and regulate a world, which had
previously seen fit to discharge them. Was ever any
pretension more intrinsically disorderly and immodest!
The apostle Paul, in the estimation of all scholars, was a
man of great sense and modesty. And the doctors
Franklin and Channing were also conspicuous for both
traits. Now is it credible for a moment that these great
men are turned into such hopeless peacocks by the mere
event of death, as to fancy that either of them is capable

of exerting the least influence upon human destiny, or
the destiny of the least individual? *Credat Judæus, non
ego.* Far easier is it for me to believe, that certain
spectral Slenders and Shallows have been donning the
dress of these good men, as found folded up and ticketed
on the shelves of somebody's reverential memory, and
vainly trying in that guise to ape also the illustrious
manners which once sanctified it.

"I am persuaded that this entire hobgoblin demon-
stration owes its existence to the superstitious and semi-
Pagan conceptions of spiritual existence which overrun
society, and which are diligently nurtured by the old
theology. The old theology represents the spiritual world
as remote from the natural one *in space*. It supposes
that when men die, they actually traverse space, actually
*go* somewhere, and bring up either at a certain fixed
*locale* within the realm of sense, constituting heaven, or
at another fixed *locale* constituting hell. Books even are
written to suggest the probable latitude of these places,
whether within or without our solar system, and so forth.
But this is clearly puerile. The spiritual world does not
fall within time and space. Time and space simply express two most general laws or methods by which the
sensuous understanding, or the intelligence enlightened
only by the senses, apprehends spiritual existence, or
gathers knowledge. Thus, man, being a creature of
infinite love and wisdom, is spiritually, or in his most
intimate self, a form of affection and intellect. But intellect and affection are purely subjective existences: they
are not *things*, visible to sense: they are forms of life.
Hence unless some plane exist, in which these forms may
be mirrored, and in which at the same time, man's

faculty may be organized to discern them, he must for ever remain unconscious of himself, devoid of conscious life. He must in fact remain for ever blent with Deity, or infinitude, and therefore dead to all that stupendous epic of passion, intellect, and action, which constitutes his present history, and which is based exclusively upon his finite natural experience.

"For nature furnishes this necessary plane, and its two universal laws, the one named time, serving sharply to discriminate to our perception event from event, and the other named space, serving sharply to discriminate to our perception form from form, supply us with the fixed alphabet of all knowledge. Accordingly whatsoever is in space and time, whatsoever falls within the realm of sense and fills the page of history, is purely phenomenal. It is not being, but only the appearance of being to a limited intelligence, an intelligence limited by the senses. Hence the sacredest incidents of history are not essential facts of humanity, but representative facts,—facts which merely symbolize infinite and eternal verities, or verities which utterly disclaim space and time. My true being, the being of every man, is God, or infinite goodness and truth. Now infinite goodness and truth, though they reveal themselves to a finite appreciation under the forms of time and space, under sensible forms, yet are not themselves sensible forms, but spiritual forms, which quite transcend time and space. Consequently my being, my essential selfhood, is always independent of space and time, and when I die therefore or become invisible to sense, the event is purely circumferential and does not affect my central quality. That remains as immutable as God, because it is God, and is consequently in no danger of being compromised by any event of my outward or sensible experience. All these events do but image, or bring to my own consciousness, the wonders of divinity which are shut up within me and in all men. And the event of death itself is only more signal than other events, because it makes this thrilling imagery more near and miraculous, by opening my consciousness to an inner field of being, in which time and space are no longer fixed but pliant to the affections of the individual, or in which every outward event and every outward form are visibly born of the subject's private selfhood, and not as here of his common nature."

---

### B.—p. 142.

The creative and eternal Word to man runs thus : "Of every tree in the garden thou mayest *freely* eat, *but of the tree of knowledge of good and evil thou shalt not eat,* BECAUSE IN THE DAY THOU EATEST THEREOF THOU SHALT SURELY DIE." Philosophers have long sought to demonstrate the reality of human freedom as evinced in the phenomena of our moral consciousness, but they have only succeeded in demonstrating the unhappy muddle Philosophy herself amounts to, so long as she superciliously disdains the guiding light of revelation, and seeks to interpret nature by the servile light which nature herself supplies. Our moral freedom is in truth only a semblance, not a reality. We *seem* to act freely, or of ourselves, when we steal or refrain from stealing, when we commit adultery or refrain from it : and man's judg-

ment accordingly, which is limited to appearances, asks
no further warrant to render us in either case blame-or-
praise-worthy. But, as Swedenborg proves on every
page of his remarkable writings, we really never do act
in freedom or of ourselves under these circumstances.
He shews by the most luminous exposition of spiritual
laws that we never steal or commit adultery, however free
the act *seem* to our foolish selves, but by the overwhelming tyranny of hell; and that we never refrain from
doing these things except by virtue of the Lord's power
constraining us to do so in spite of our natural tendencies.
We *feel* this power to be in ourselves, that is to be freely
exerted, only because we do not sensibly discern the fields
of spiritual existence from which alone it inflows, and
our senses have hitherto ruled our reason in place of
serving it. No man since the world has stood has ever
had power to draw a physical or moral breath, independently of those celestial and infernal companies with
which all his past ancestry interiorly but unconsciously
associates him. Of course therefore the Divine Love is
incapable of ascribing any one's physical and moral
merit or demerit to the person himself, because it would
be absurdly false to do so. On the contrary, it seeks
with endless pains to prevent the man himself from doing
this by the organization of conscience as an unfaltering
ministry of death. Our most accredited theologies and
philosophies have always alike misapprehended the scope
of this relentless ministry. They suppose that conscience
was originally intended as a ministry of life or righteousness, and that Adam accordingly enjoyed its favourable testimony in Paradise before he had eaten of the tree
of knowledge, that is, before he had learned to appro-

priate good and evil to himself. But of what possible use could the approbation of conscience be to a being who was still ignorant of the difference between good and evil ? The transparent contradiction involved in the assumption sufficiently demonstrates its absurdity to the reason; but the literal text of revelation demonstrates it also to the very senses, by shewing us that conscience first dawned in Adam after self hood (Eve) had been developed in him, and he had been led by it to eat of the tree of knowledge, that is, to appropriate his influent good and evil to himself.

Adam symbolizes the immature condition of the mind, the merely seeming and constitutional side of man, the life of instinct which we derive from nature, and which through the decease operated in us by conscience, we ultimately lay aside in order to the assumption of our true and spontaneous life derived directly from God. Hence —what is perfectly consistent, if you regard the spiritual purport of the narrative, but perfectly absurd if you regard only its letter—the most pregnant service which Eve (representing the divinely endowed self hood) renders Adam, is to throw him instantly out of Paradise, by unmuzzling within him the relentless jaws of conscience. Do you ask me what I mean by Eve, as the symbol of our divinely quickened self hood? I will tell you. Adam, as we have seen, represents our finite or constitutional existence, that which flows from our connection with the race. It seems to be a most real existence, while in truth it is a purely reflected one, the subject being nothing but what he is made by the spiritual world, being in fact as destitute of real self hood or freedom as if he were only dove or rabbit and not man. The dove or rabbit remains spiritually unquickened, devoid of true individuality, because it is a purely animal form, that is, a form in which the universal element dominates the individual one.   It is, in other words, and ever remains, an unshrinking subject of its nature, and hence incapable of " eating of the tree of knowledge of good and evil," that is, of appropriating good and evil to itself.   It has no consciousness of a selfhood underived from its nature, and is consequently utterly incapable both of moral experience, and of that lustrous life of conjunction with God in which such experience, when left unperverted, infallibly merges. But Adam, the beautiful symbol of our nascent humanity, of our still instinctual and *pre-moral* beginnings, is lifted above animality by his human form, that form being the only one in which the universal element *serves* the individual one, and which therefore fitly images God. Hence, though he is but a rudimental and seeming man, he is bound at once to vindicate his essential divinity, by exhibiting, in however rude and purely negative a form, the real and distinctive life which animates humanity.   The distinction of man from all lower existences is, that he is in strictest truth the child of God, that Infinite Love and Wisdom constitute his veritable and exclusive parentage, and Infinite Love and Wisdom are utterly inconsistent with selfishness or with littleness of any description. Hence in Christ, who is the perfected and fully conscious Divine Man, we see his merely finite and *quasi* or constitutional life, his purely Adamic selfhood, incessantly deposed in order to his glorification, in order to his consummate union with God.   In Adam consequently, who is but the prophetic or typical and unconscious divine man, we must expect to see death installed as the very *fons et principium vitæ,* as the very fountain and spring of human life; we must expect to see despair enthroned as the fertile and abounding womb of man's distinctive hope. In short, we must demand from Adam, as the symbol of our rudimental and initiatory manhood, a purely negative and mortuary experience; that is to say, we must expect to see him divorced from his merely seeming and dramatic existence, by falling under the dominion of conscience or the moral law.

By Eve, then, or our divinely vivified selfhood, is meant the power which is incessantly communicated to man of separating himself from his mere animal conditions, of elevating himself out of the realm of law into that of life, or of subjecting nature and society to the needs of his individuality. In short, Eve signifies the power in man of spiritually appropriating good and evil to himself, the faculty of spiritual consciousness. Let the animal do as he will, he is but the abject vassal of his nature, and therefore destitute of personality or character, destitute of spiritual consciousness. The animal is only naturally, never spiritually, good or evil. The dove is naturally good as contrasted with the vulture, the tiger is naturally evil as contrasted with the sheep, but you would never think of deeming the dove spiritually good or the tiger spiritually evil as contrasted with any other animal, especially as contrasted with any other dove or tiger. Why? Because spiritual good and evil is individual good and evil, that is, it implies in the subject a spiritual individuality uncontrolled by his nature. The animals have no such individuality, and hence are ignorant of moral distinctions, are unworthy of individual praise or blame. They have a purely natural individuality, and hence are incapable of eating of the tree of knowledge of good and evil, or of viewing themselves as spiritually responsible for their influent good and evil. Of man alone is it lawful to predicate moral distinctions, because he alone is capable of appropriating his influent good and evil to himself in place of charging it upon his nature. He alone is capable of an alternate individual expansion and collapse, which unless the Divine Mercy overruled them to his endless benefit, would breed only the most disastrous consequences. He is capable at one moment of a spiritual conceit and pride which plunges him gaily into hell, at the next of a spiritual despair which shuts him sorrowfully out of heaven. For example, if moral good prevail in my natural disposition, if I pass my life in visiting prisons, building hospitals, feeding the poor, scattering tracts, circulating the Bible, forwarding every conventionally righteous enterprize, while maintaining at the same time an irreproachable private and social deportment, I shall be infallibly certain —unless the Divine Love expose me to incessant secret or spiritual shipwreck, to the most withering internal humiliation and disaster—to appropriate this good to myself, and so turn out a monster of spiritual pride, a being too inflated even for hell to tolerate. Or if moral evil preponderate in my natural character, if on all occasions of temptation I succumb, and convict myself of lying, theft, adultery, and what not, I shall be sure in these circumstances—unless the Divine Love visit me with incessant outward success and prosperity—to shut myself up in a despair too obdurate even for the warmest love of heaven to penetrate it. These experiences, mournful as they seem when too narrowly viewed, nevertheless

attest the grandeur of human nature.  They are possible
to us only because our distinctively human life dates from
God, and is therefore a spontaneous life, a life whose
principle of action falls exclusively within the subject,
and renders him therefore eternally free.  Of course this
life presupposes the complete reconciliation of self-love
with brotherly love, presupposes the scientific inauguration of human society, human fellowship, human equality,
and these issues again presuppose a conflict of these two
forces, suppose, that is, a previous stage of human experience in which self-love is at war with brotherly love,
or hell antagonizes heaven in lieu of promoting it.  Now
so long as this infantile state of things endures, so long
as self-love and brotherly love, or hell and heaven, are
kept unreconciled by the immaturity of the scientific understanding in man, we each of us, by virtue of the
solidarity that binds us to the race, feel this conflict in
our own bosoms as if it originated there, or belonged to
ourselves, instead of being a veritable influx from the
entire spiritual world, or the universal mind of man.
We have not the least suspicion that the conflict is not
our own private affair, is not a legitimate feature of our
divinely given individuality, and accordingly as one or the
other principle prevails in our life, we contentedly write
ourselves down good or evil in the Divine sight, turning
out wretched Pharisees in the former case, and despised
publicans and harlots in the latter.  We have not the
slightest conception of our true and spontaneous life, nor
consequently of the miraculous exhibitions of Divine
wealth and power with which it is fraught.  We have no
idea that that life is so divinely majestic and perfect as to
involve in itself the complete reconciliation of hell and

heaven, the intensest harmony of self-love and brotherly love, of the external and internal man. Not knowing this, we inevitably suppose that our spiritual experience belongs to our isolated private bosoms, qualifying us individually in the sight of God; and we therefore go on to eat of the tree of knowledge of good and evil with a stupid *gusto* that of necessity disallows the true consciousness of God in our souls, and turns His inward voice of love and mercy into one of implacable condemnation and death.

---

### *C.*—p. 159.

"THE INCLINATION TO UNITE THE MAN TO HERSELF IS CONSTANT AND PERPETUAL WITH THE WIFE, BUT INCONSTANT AND ALTERNATE WITH THE MAN. The reason of this is, because love cannot do otherwise than love and unite itself, in order that it may be loved in return, this being its very essence and life; and women are born loves; whereas men, with whom they unite themselves in order that they may be loved in return, are receptions. Moreover love is continually efficient; being like heat, flame, and fire, which perish if their efficiency is checked. Hence the inclination to unite the man to herself is constant and perpetual with the wife: but a similar inclination does not operate with the man towards the wife, because the man is not love, but only a recipient of love; and as a state of reception is absent or present according to intruding cares, and to the varying presence or absence of heat in the mind, as derived from

various causes, and also according to the increase and
decrease of the bodily powers, which do not return regu-
larly and at stated periods, it follows, that the inclination
to conjunction is inconstant and alternate with men.

"CONJUNCTION IS INSPIRED INTO THE MAN FROM
THE WIFE ACCORDING TO HER LOVE, AND IS RECEIVED
BY THE MAN ACCORDING TO HIS WISDOM. That love
and consequent conjunction is inspired into the man by
the wife, is at this day concealed from the men; yea, it is
universally denied by them; because wives insinuate that
the men alone love, and that they themselves receive; or
that the men are loves, and themselves obediences: they
rejoice also in heart when the men believe it to be so.
There are several reasons why they endeavour to persuade
the men of this, which are all grounded in their prudence
and circumspection. The reason why men receive from
their wives the inspiration or insinuation of love, is, because nothing of conjugial love, or even of the love of the
sex, is with the men, but only with wives and females.
That this is the case, has been clearly shewn me in the
spiritual world. I was once engaged in conversation
there on this subject; and the men, in consequence of a
persuasion infused from their wives, insisted that they
loved and not the wives; but that the wives received love
from them. In order to settle the dispute respecting this
arcanum, all the females, married and unmarried, were
withdrawn from the men, and at the same time the sphere
of the love of the sex was removed with them. On the
removal of this sphere the men were reduced to a very
unusual state, such as they had never before perceived,
at which they greatly complained. Then, while they were
in this state, the females were brought to them, and the

wives to the husbands; and both the wives and the other females addressed them in the tenderest and most engaging manner; but they were cold to their tenderness, and turned away, and said one to another, "What is all this? what is a female?" And when some of the women said that they were their wives, they replied, "What is a wife? we do not know you." But when the wives began to be grieved at this absolutely cold indifference of the men, and some of them to shed tears, the sphere of the love of the female sex, and the conjugial sphere, which had for a time been withdrawn from the men, was restored; and then the men instantly returned into their former state, the lovers of marriage into their state, and the lovers of the sex into theirs. Thus the men were convinced, that nothing of conjugial love, or even of the love of the sex, resides with them, but only with the wives and females. Nevertheless, the wives afterwards, from their prudence, induced the men to believe that love resides with the men, and that some small spark of it may pass from them into the wives."—*Conjugial Love*, n. 160, 161.

----

### *D.—p. 229.*

It is of this historically avouched decline of moral power in humanity that my admired friend Carlyle complains in so many exquisite pages of mingled pathos and invective. Carlyle has apparently not the slightest conception of the new and perfect manhood which is dawning, and cherishes every vestige of the old forceful and fanatic type in that sort, as *Old Mortality* cherished the fading *hic jacets* upon the tombstones of the martyrs. Carlyle's heroes no doubt were a good style of men in their day, but their day was strictly in order to ours, which, bemoan it as you will, is an incomparably brighter one for humanity than the earth has ever before known. What should we say of a gardener who went on cultivating the gnarled and sturdy trunk of a vine, long after it had yielded all the fruit it was capable of? Precisely the same must we think of Carlyle, whose infatuation consists, not in the desire reverently to bury the past, but to revive it in conditions which would be obviously and utterly fatal to its continued existence for a moment. The truth is, Carlyle is an accomplished artist, who hangs one's house with historic portraits and *tableaux* of an incomparable lustre; but as to the scope of history itself—as to what men call the philosophy of history, meaning thereby the great human soul which gives it the unity of a man, and which is fast coming to superb and perfect consciousness by it— he is so exquisitely blind as to be even scornfully vituperative. It would make your ears tingle to hear the thunderous mirth with which on occasion he belabours the scientific conception of human destiny, to hear the great guns of riotous laughter which he lets off in broadsides upon the poor innocent soul who fancies that a science of history is strictly possible—he who will abide no other ideal for man than that of proving an eternal bruiser. I cannot help, much as I esteem Carlyle, recognizing here the essentially Barnum conception of manhood, never unconscious youthful grace and symmetry, but everywhere gigantic overgrowth contrasted by dwarfish undergrowth. The Barnum type of godhead is strictly proportionate: nowhere the benignant power which gives life to all things by sedulously concealing itself and shunning recognition; but some egregious posture-master, who on a set day plants himself in the centre of space and dramatically conjures all things out of nothing in a way to astonish Robert Houdin himself, and all whom he astonishes. But let Carlyle pipe what melodious notes he pleases (and surely I shall be the last to grow weary of listening), science is bent upon utterly sapping our reverence for the great historic names, simply because this greatness all proceeds upon the implication of will or moral force, and science traces all will, all morality, to the devil, that is, to the servile side of human nature. Morality in fact is only the Divine method of taming the devil, or schooling him to the hearty allegiance of man. Moral life is born of an enforced subjection of the affections to the intellect, of the individual sentiment to the common one; whilst the Divine or spontaneous life is born of a cordial marriage between the affections and the intellect, between the sentiment of individuality and that of community. Hence it is, that so long as the moral régime endures, so long as men continue to eat of the accursed fruit "of the tree of knowledge of good and evil," or foolishly appropriate to themselves what science declares to be exclusively from angelic and infernal association, so long the Divine voice in their souls must prove a ministry of death, and jealously obstruct the way of the tree of life. Carlyle's Cæsars, Mahomets, Cromwells, Napoleons, were above all things men of a defective spiritual fibre; in other words, their natural vigour is so much beyond the sane average, that they either obdurately resist this Divine voice altogether, or else dexterously pervert it to the authentication of their fanatical self-conceit, in which case Providence

kindly adopts and tickets them as so many hardy and
consummate policemen, fit to dragoon humanity into a
temporary semblance of order.  But the idea of confounding for a moment any policeman, much more any eccle-
siastic, with a man, with God's finished work!—surely
this is unworthy of Carlyle, and leaves his genius—not
indeed his rhetorical, but his real genius—far below that
of many men who will never make half his noise.  Heaven knows that no one would quarrel with hero worship
as a feature of human history, as a passing manifestation
of the immortal cultus for which the human heart is primarily constructed.  What one quarrels with is to see a
grave sincere soul like Carlyle pining for the restoration
of that sort of thing, helplessly incapable of extracting
its majestic human meaning for it once for all, and so
bidding it a jolly good-bye for ever.  He seems at moments to have a glimpse of their being something symbolic
in hero-worship, and yet he neglects the first law of
symbolic hermeneutics, which is, that the symbol occupy
a lower plane than the thing symbolized: for he evidently
regards hero-worship in the past as prophetic only of continued hero-worship in the future: that is, he makes the
symbol symbolize itself.  But all this is absurd.  The
symbol always stands for something it cannot comprehend,
something which cannot be transacted in the same region
with itself.  Thus viewed, hero-worship does not mean
any such inconsequence as that stupid people are for ever
going to gaze open-mouthed on clever Divine corporals
and lieutenants occasionally raised up to further Providential ends: it means that human nature itself is to become
so transfigured by its divine head, as that every man, by
virtue merely of his human form, will be a conspicuous

temple of Deity, and that homage consequently which we now render to the creature be turned into a cordial and joyous tribute only to manifested Deity.

---

### E.—p. 232.

I do not complain of course that the inseparable distinction of good and evil is made too much of. No man, not an idiot, can ever fail to abhor lying, theft, adultery, and murder, as features of human conduct, nor consequently to applaud the habitual and scrupulous abnegation of these things; because our spiritual existence is conditioned upon that bipolarity, just as our physical existence is conditioned upon the bipolarity of pleasure and pain; and to suppose one indifferent therefore to moral distinctions is to suppose him spiritually non-existent, just as to suppose one indifferent to the distinction of pleasure and pain is to suppose him physically non-existent. In both spheres alike these things are the mere constitutional conditions of our existence, and what I quarrel with consequently is that they should not be left in that intensely subordinate plight, but become exalted by our foolish theologians and philosophers into the very sources also of our life. My animal consciousness is constituted by my susceptibility to pleasure and pain, or my relations to outlying nature; and this consciousness would for ever immerse me as it does the horse or the tiger, and prevent my rise to the human level, were it not for conscience acquainting me with a superior pleasure and a profounder pain, and so rescuing me from its grasp.

But if I hereupon insist upon identifying myself with these new conditions of existence, if I insist upon confounding my proper life in this new sphere with the mere organization which serves to develope it or give it manifestation, I shall practically incur the same mistake as the man who makes freedom to mean no diviner thing than emancipation from fetters, and shall remain, under my moral tutelage, even more hopelessly remote from true communion with God than I had been before as a simple animal. The animal existence is never diabolic. The human is invariably so during its transition from instinct to spontaneity, or while truth in the intellect instead of good in the heart rules the conduct. " The reason of man," says Swedenborg, *Arcana Cœlestia*, 1949 and 1950, "is made up of good in the heart and truth in the understanding. Good is the interior celestial element and constitutes the very soul or life of the reason: truth is the exterior spiritual element and is what receives life from that interior good. Rational truth uninspired by good is symbolized by Ishmael: it fights against all, and all fight against it. Rational good never fights, howsoever it is assailed, because it is meek and gentle, patient and accommodating, all its attributes being those of love and mercy: and although it does not fight, yet it conquers all, never thinking of combat nor boasting of victory. It acts thus because it is divine, and is safe of itself: for no evil can assault good, nor even subsist in the same region. If it feel even the approach of good, evil spontaneously recedes and retires. And what is true of good, is true in a measure, also, of truth enlivened by good, because such truth is only good in form. But truth separate from good, or unvivified by it, which is represented by Ishmael, is of a different quality. It thinks and devises scarcely anything but combats, its ruling affection being to conquer, and when it conquers it boasts of its prowess. A man of this sort, though he be in the most orthodox truth of faith, if he be not at the same time animated by charity, is morose, impatient, querulous towards all the world, viewing every body else as in error, zealously rebuking, chastising, correcting; he is without pity, nor does he endeavour tenderly to bend the affections and thoughts of others to what he conceives to be right, [but on the contrary seeks to *coerce* them into his own way of thinking :] for he regards everything from the point of view of truth, nothing from the point of view of good." And yet this fine-hearted and deep-thoughted old man is of course pronounced insane by every theologic or philosophic noodle in the land.

---

### ERRATA.

p. 32, l. 13 from bottom, insert a colon after *knowledge*.
p. 85, last line, for *uncreated*, read *unformed*.
p. 86, l. 12 from top, strike out the second *and*.
p. 124, l. 16 from top, for *is*, read *are*.

Mitchell and Son, Printers, Wardour Street (W.)

CPSIA information can be obtained
at www.ICGtesting.com
Printed in the USA
BVHW041622110619
550711BV00014B/240/P